**Cristian Jiménez-Martínez**
**Milagros Faridy Juárez-Chairez**
**María Stephanie Cid-Gallegos**
Editors

# Legumes: Beyond Their Nutritional Value

nova
science publishers
www.novapublishers.com

DOI: https://doi.org/10.52305/IEHX2115



**Library of Congress Cataloging-in-Publication Data**

ISBN: 979-8-89530-440-2 (Hardcover)
ISBN: 979-8-89530-498-3 (eBook)

*Published by Nova Science Publishers, Inc. † New York*

# Contents

# Preface

Legumes, a diverse group of plants within the Fabaceae family, have been a cornerstone of human nutrition and agriculture for millennia. With a rich history dating back to 7000-8000 B.C. in regions like Turkey, legumes were among the first plants to be domesticated by humans. Today, they remain a crucial dietary component worldwide, particularly for those who choose plant-based diets or lack access to animal-based foods. Beyond their nutritional value, legumes such as beans, chickpeas, soybeans, and peas (although the *Fabaceae* family comprises over 19,400 species, some of which are consumed locally) have long been valued in traditional medicine for their versatile therapeutic properties, including analgesic, diuretic, anti-inflammatory, and antibiotic effects. Modern scientific research has further highlighted the bioactive compounds present in legumes, such as phenolic compounds, peptides, saponins, and isoflavones. These compounds contribute to legumes' health benefits, including antioxidant, anticancer, and antidiabetic properties, and play a vital role in the prevention and management of various chronic diseases. As global populations and demands for sustainable food sources increase, understanding the nutritional and therapeutic potential of legumes is more important than ever. In these chapters aims to explore the chemical composition of various legumes (traditionally and locally consumed), their bioactive compounds, and the impact of different processing techniques on these compounds. We will examine how these elements contribute to human health, considering their potential applications in biotechnology and the development of functional foods.

# Acknowledgments

Editors are grateful to the authors for the excellent quality of the chapters, the chapter manuscript's reviewers, and everyone who has worked on this book to get it published. Thanks to Nova Science Publishers for making the publication of this book possible and to the National Polytechnic Institute for its support.

## Chapter 1

# Beneficial Outcomes of Bioactive Compounds from Legumes

**M. F. Juárez-Chairez**[1]
**M. S. Cid-Gallegos**[2]
**Y. M. Gómez y Gómez**[2]
**L. J. Corzo-Ríos**[2]
**X. M. Sánchez-Chino**[3]
**and C. Jiménez-Martínez**[4,*]

[1]Laboratorios de Inmunología y Biología Molecular, Unidad Académica de Ciencias Biológicas de la Universidad Autónoma de Zacatecas, Guadalupe, Zacatecas, México
[2]Unidad Profesional Interdisciplinaria de Biotecnología, Instituto Politécnico Nacional, Mexico City, Mexico
[3]CONAHCyT, Departamento de Salud, El Colegio de la Frontera Sur-Villahermosa, Villahermosa, Tabasco, Mexico
[4]Escuela Nacional de Ciencias Biológicas, Instituto Politécnico Nacional, Unidad Profesional Adolfo López Mateos, Mexico City, Mexico

## Abstract

Legumes, members of the Fabaceae family, encompass approximately 650 genera and 18,000 species. They are among the earliest plants domesticated and cultivated by humans, with evidence of their use in Turkey dating back to 7000-8000 B.C. They are also regarded as an essential source of protein for human consumption, particularly for individuals who follow a plant-based diet. It is important to note that since ancient times, beyond their use as food to meet nutritional needs,

---

* Corresponding Author's Email: crisjm_99@yahoo.com

In: Legumes: Beyond Their Nutritional Value
Editors: C. Jiménez-Martínez, M. F. Juárez-Chairez and M. S. Cid-Gallegos
ISBN: 979-8-89530-440-2

the seeds, leaves, and roots of legumes have been utilized in traditional and folk medicine by indigenous peoples for their analgesic, antidiarrheal, diuretic, laxative, antibiotic, and anti-inflammatory properties, among others. Within conventional medicine, legumes such as beans, chickpeas, soybeans, and peas have been subjected to various treatments or extraction methods allowing them to be used for medicinal purposes. Hence, it is important to know the origin and development of legumes in traditional medicine to understand better how these seeds' bioactive compounds will benefit health.

**Keywords:** traditional medicine, legumes, bioactive compounds, health

## 1. Introduction

Traditional or folk medicine is a cultural practice that has been carried out since ancient times, including treating or ceasing diseases. Humanity has accumulated knowledge for generations about plants with medicinal effects on specific diseases (Ogu, 2022). Additionally, the World Health Organization (WHO, 2023) refers to traditional medicine as the set of all knowledge, skills, practices, and beliefs, whether or not explained. They are used to prevent, diagnose, and improve physical and mental health.

The Egyptian civilization gathered information about the medicinal properties of plants. They identified several compounds in papyri recommended for various diseases found in pyramids and tombs. For example, the first medical treatise of this civilization was found in the "Papyrus Ebers," with more than 800 medicinal preparations, which dates from 1500 BC and is considered the oldest collection (Polito et al., 2016). Additionally, flowers, seeds, fruits, and vegetables were subjected to heat treatments (boiling, roasting, etc.) or extractions with water or organic solvents and reported in treatises on medicinal plants of some ancient civilizations. Thus, modern/western medicine has evolved from traditional medicine. Currently, in certain parts of the world, knowledge of medicinal plants is the only option available for medical treatment (Phumthum et al., 2020).

In addition, ethnopharmacological knowledge in Europe has its roots in Greek and Roman culture, where Dioscorides, Pliny the Elder, Galen, Theophrastus, and Hippocrates exerted a significant influence, and herbalism played a major role in the discovery of medicines in the early XIX century (Petran et al., 2020).

Regarding seeds or plants attributed with medicinal properties, we could mention legumes, the third largest group of flowering plants (Borges et al., 2013; Didinger & Thompson, 2021); 88% of them can form nodules with rhizobia, which allows them to fix nitrogen in the soil and promote plant growth (Simonsen et al., 2017). Domestication of these plants occurred primarily in Asia and the Americas during different historical periods. For instance, lentils have been traced back to between 9500 and 8000 B.C. at a site in Iran. Additionally, beans and soybeans were domesticated over 3,000 years ago in the Americas and Asia, respectively (Graham and Vance, 2003).

## 2. Legumes Through History

Legumes have played a significant role in human history, being one of the most important sources of protein of plant origin. Additionally, over the years, they have positioned themselves as a priority to reduce protein consumption from animal sources (Duranti, 2006). Moreover, beyond their use as food, they are considered to have biological properties of interest for preventing non-communicable chronic diseases (NCCD) (Giangrieco et al., 2022).

The advent of agriculture marked a before and after in the evolution of humanity, driving cultural development, biodiversity, human health, and rapid population demographic expansion. The "Neolithic Revolution" also saw the emergence of new stone and ceramic tools used primarily for processing starchy foods, storing milk or beer, and storing crop surpluses and products, leading to a sedentary lifestyle (Craig et al., 2011). In addition, isotope analysis has shown that Neanderthals consumed protein-rich diets from animal and plant products, as isotope analysis only records the nitrogen in the diet (Henry et al., 2010).

Thus, the consumption of these seeds dates to the Neolithic period, when various plants were domesticated (Cristiani et al., 2016). These seeds were found in charred spaces intended for fires or hearths during the Neolithic period between 7000 and 8000 BC in Turkey, just as soybeans began to be cultivated in China between 2000 and 3000 BC. However, information on whether Neanderthals cooked their food is scarce (Henry et al., 2010). In addition, beans, soybeans, and other crops in America and Asia were domesticated more than 3000 years ago and became staples of various cultures (Ahmed & Hassan, 2014).

Furthermore, it is essential to highlight that the domestication of legumes, along with the use of tools such as ceramic vessels, enabled various treatments to be applied to the seeds. One such treatment is soaking, which involves bringing the seeds below the gelatinization temperature while increasing their moisture content to shorten cooking time. Subsequently, the seeds are heated above the gelatinization temperature of starch to produce a soft, flavorful product (Güzel & Sayar, 2012).

Besides, they discovered various cures for specific common ailments by eating or making aqueous extracts from the plant as a whole or a specific part. Hence, they began to use the plant in folk medicine to prevent or cure certain diseases (Figure 1). The biological effect depended on the seeds (chickpea, bean, soybean, pea) and the geographical region where they were found (Figure 1) (Ahmed & Hassan, 2014).

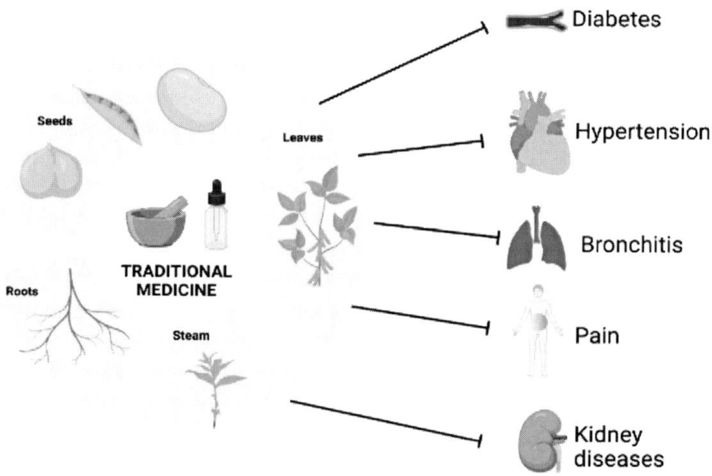


**Figure 1.** Traditional medicine plants and applications. Created in Biorender.

## 3. Medicinal Uses of *Phaseolus vulgaris*

*Phaseolus vulgaris* is one of the most consumed legumes in the world due to its essential macro and micronutrient content (Hayat et al., 2014). Moreover, Mexico is where the domestication of the common bean began, and as it spread throughout South America, two eco-geographically distinct gene pools emerged (Bitocchi et al., 2013). In addition to its use as a food, it has also been used in traditional medicine, where it has been attributed with various properties, such as analgesic for back and knee pain, diuretic, and effective for diarrhea, amebiasis, conjunctivitis, lower extremity edema, hypertension, intestinal ulcers, painful urination, burns, mumps, and food poisoning (Ahmed & Hassan, 2014). Additionally, antidiabetic properties have been attributed to beans since the 1920s, when ethanolic or aqueous extracts of the pods were used, which have been observed to lower blood glucose in diabetic patients (Helmstädter, 2010).

A study conducted in the central region of the Republic of Benin analyzed the leaves, seeds, and roots of various bean plants (*Phaseolus vulgaris*) used in traditional medicine in the region (Loko et al., 2018), finding that the seeds were used for health conditions such as pharyngitis and to accelerate the healing process of the umbilical cord. In addition, the leaves and roots were used in infusions or mixed in water to treat wounds, bad body odor, bee stings, vaginal infections, sterility, fever, and obesity. Moreover, Bljajic et al., (2021) studied extracts of Croatian ethnomedicinal plants and found that the aqueous bean extract had antioxidant activity and inhibited the oxidation of β-carotene. Beans represent an alternative to conventional medicine to treat various diseases (Bljajic et al., 2021).

However, Damián-Medina et al. (2020) found in an *in silico* study that polyphenols present in beans can modulate the activity of proteins in type 2 diabetes, signaling pathways such as insulin secretion, insulin resistance, and carbohydrate absorption. For this reason, they suggested that foods rich in polyphenols can be used to prevent metabolic diseases. Similarly, Lomas-Soria et al. (2015) conducted a study with three varieties of cooked P. vulgaris like Flor de Mayo bean (FMB), Peruan bean (PB), and Negro bean (NB) in a streptozotocin-induced diabetic rat model. They observed that after 4 weeks of administering the FMB variety, glucose levels decreased and insulin levels increased, which was attributed to the synergistic effect of the compounds in the bean. In addition, P. vulgaris has been used as a medicinal plant in Europe for its anti-diabetic effect attributed to α-amylase inhibitors; it can control type 2 diabetes by inhibiting DPP-IV (Tran et al., 2020; Przeor, 2022).

## 4. Medicinal Uses of *Glycine max*

Soybean (*Glycine max*) belongs to the *Leguminosae* family and grows in tropical, subtropical, and temperate climates (Nyandwaro et al., 2020). It is a significant crop due to its oil and protein content; its seeds contain 20% oil (unsaturated fatty acids such as oleic acid and linoleic acid) and is rich in bioactive compounds important for health (FAO, 2023; Nyandwaro et al., 2020). Its origin is in China and East Asia, although today, 80-85% of its cultivation is in the Western Hemisphere, with Brazil and the United States being the leading producers (Anderson et al., 2019).

In addition, soybean has components with biological activity for human health that have fostered scientific interest in its analysis since ancient times; traditional medicine from China, Japan, Korea, and other countries have attributed healing properties to it (Lee et al., 2020). For example, soybean leaves, roots, and seeds have been used to treat diseases such as malaria, atherosclerosis, osteoporosis, and several types of cancer such as prostate, breast, and uterine. Antioxidant activity is also attributed to compounds of phenolic origin or nature, such as isoflavones (Nyandwaro et al., 2020). Additionally, isoflavones are present in large amounts in soybeans, where they predominate in the form of genistein and daidzein conjugated to various sugars in the form of malonyglucosides, acetylglucosides, and glycosides, which are present in a complex and varied composition (Setchell, 2017).

Moreover, genistein and daidzein are similar to estradiol found in mammals and exert antioxidant activity both *in vitro* and *in vivo* by increasing hepatic cytosolic glutathione peroxidase activity (Ahmad et al., 2014). Therefore, they are used in traditional medicine to treat various aging diseases such as neurodegenerative disorders, osteoporosis, cardiovascular diseases, and symptoms of menopause, observing that the effects of menopause improved in women who consumed soy foods (Jung et al., 2020).

Similarly, Sugano (2006) mentions that other compounds found in soybeans, such as α-linoleic acid, lecithins, lectins, linoleic acid, proteins, phytosterols, and saponins, may have certain health-promoting biological activities, like lowering cholesterol and triglycerides and improving memory and learning ability.

Additionally, a study was conducted using methanolic extracts and peptides from soybean (*G. max*) administered for 4 days (800 mg/kg) in an in vivo *Plasmodium berghei*- malaria mouse model. The results showed a reduction in mean parasitemia levels to3.48 ± 0.37 (methanol) and 3.61 ± 0.27 (peptide), corresponding to a 73% decrease compared to the positive control (Nyandwaro et al., 2020).

## 5. Medicinal Uses of Chickpea (*Cicer arietinum* L.)

Chickpea is one of the most consumed legumes, especially in the Mediterranean and the Western world (Gupta et al., 2016). These immature beans can be eaten raw, and the ripe ones can be dried and ground into flour and used as animal feed or as a coffee substitute. Cooked, they are added to salads, especially in Western Europe and the USA, whereas in the Middle East, they are consumed surrounded by sugar or spices; in addition, chickpea flour is also used as a complement to make pasta, soups, and bread (Rachwat-Rosiak et al., 2015). Likewise, chickpea is considered a healthy food due to its dietary values, therapeutic characteristics, and health benefits as it has been used in traditional medicines, like Persian, Uyghur, Indian, Chinese, and Roman (Kumar et al., 2022). Persian medicine is among the oldest and most valuable traditional systems of medicine, in which chickpea broth (Nokhodab), a conventional Persian food widely consumed in Asia containing chickpeas, meat, rice, saffron, onion, and cinnamon, has been used as a traditional treatment of osteoarthritis for many centuries, to treat weakness of the body, and in diseases caused by abnormal accumulation of phlegm or yellow bile (Ahmadi et al., 2020; Heydarirad et al., 2019).

Similarly, chickpea was prescribed as a perfect medicinal food for lung enlargement. It was used to treat lung ulcers; taken orally with vinegar, it was considered highly effective in killing intestinal worms or parasites and an important aphrodisiac agent. Besides, it was prescribed to eat chickpeas soaked in water with a little honey for patients with impotence. The black-skinned chickpea (Desi type) was known more as a medicine than food and was recommended as an abortifacient, antidote, kidney stone crusher, and diuretic agent (Memariani et al., 2020).

In traditional Persian medicine, chickpeas were prescribed for cancer and believed to enhance libido and stimulate breast milk production, while chickpea oil was used to alleviate joint pain (Abbasifard et al., 2020; Kumar et al., 2022). In traditional Uyghur medicine, chickpeas have been employed for the treatment and prevention of numerous diseases, particularly type 2 diabetes, hyperlipidemia, hypertension, bronchitis, and coprostasis (Aisa et al., 2019; Kumar et al., 2022). It has been reported to have the function of removing abnormal body fluids and opening the blockage of body fluids, and used to treat body weakness, loss of appetite, itchy skin, flatulence, low libido, osteoporosis, and tumor formation (Li et al., 2021; Memariani et al., 2020). In Ayurvedic medicine, chickpea is used to treat obesity, blood dyscrasias, ear infections, disorders of the liver, gallbladder, and spleen, ailments, throat problems, bronchitis, disorders of the blood, skin, and other organs (Katiyar & Mishra, 2015; Kumar et al., 2022; Memariani et al., 2020). In Egypt, chickpea beans have increased body weight and cured headaches, sore throats, and coughs. Likewise, powdered grains are used to prepare facial masks and are added to anti-dandruff products (Rachwat-Rosiak et al., 2015). The

Romans noted that chickpea seed or leaf preparations were beneficial in treating menstrual and urinary disorders, warts, jaundice, and soothing gout pains. In India, the acidic secretions of chickpea leaves are used in traditional preparations to treat indigestion and constipation (Kumar et al., 2022).

Chinese medicine has been used to treat hypertension and diabetes for the last 2500 years (Memariani et al., 2020; Wei et al., 2017). Other traditional medicinal uses of chickpeas are against ringworm, to treat kidney stones and nephralgia, to expel parasites from the body, quench thirst, and reduce the burning sensation in the stomach (Katiyar & Mishra, 2015; Kizilarslan-Hancer et al., 2019), aside from inducing menstruation, accelerating labor, and stimulating lactation, Finally, chickpea leaves, in traditional medicine, have astringent properties; once cooked, they can be applied to displaced bones and dislocations, whereas the extract can be used as a treatment for diarrhea or indigestion (Ali et al., 2021; Gupta et al., 2016).

## 6. Medicinal Uses of Other Legumes

Traditional medicine is the oldest form of healthcare in the world and is used to prevent and treat physical and mental illnesses (Abbasifard et al., 2020). Within traditional medicine, we can find other legumes, not as popular as beans, soybeans, and chickpeas, that are also used because they are accessible, economical, and natural, which will be described in the following.

### 6.1. Pigeon Pea (*Cajanus cajan*)

The pigeon pea is a plant cultivated mainly in Asia and Africa that has been used as a food and medicinal plant. In traditional Chinese medicine, it is indicated for pain relief and as a sedative. Likewise, pigeon pea leaves are used to stop bleeding, as an analgesic, and kill parasites (Pal et al., 2011). In the case of traditional Indian medicine, pigeon pea leaves are used to treat diseases such as diabetes, hepatitis, and measles (Cui et al., 2015; Dinore & Farooqui, 2020). Likewise, the leaves, seeds, and young stems are used to cure gingivitis stomatitis and as a toothbrush. In addition, fresh juice or boiled leaves are administered orally to nullify the poisoning effect. The leaf paste is also applied to ulcers and oral inflammations as a laxative.

The leaves and seeds are applied as a poultice on the breast to induce lactation (Pal et al., 2011; Tungmunnithum & Hano, 2020). Other uses given to the pigeon pea seed in different traditional medicines are in the treatment of diarrhea, gonorrhea, chickenpox, hepatitis, constipation, and superficial infections, and to stabilize the menstrual period and in food poisoning. The scorched seeds are added to coffee to relieve headaches and vertigo. In addition, fresh seeds are believed to help with urinary incontinence in men, whereas immature seeds are recommended for treating kidney diseases (Cui et al., 2015; Tiwari et al., 2013). Likewise, the leaves are used for burns, eye infections, earaches, sore throats, gum pain, toothaches, anemia, intestinal worms, dizziness, epilepsy, sores, wounds, abdominal tumors, diabetes, malaria, bedsores, as well as treatment against hypoxic-ischemic brain damage and alcohol-induced liver damage. The young leaves of the pigeon pea are chewed to treat coughs, diarrhea, trauma, burn infections, bedsores, toothaches, sore gums, childbirth, and dysentery. In the case of root preparations, they are taken to treat coughs, stomach problems, and syphilis. The dried roots are used as anthelmintic, expectorant, and sedative. Stem ashes are used for wounds, and the stems and roots are chewed against toothaches. Finally, the powdered seeds serve as a poultice on swellings (Sahu et al., 2014; Srivastava, 2013; Tiwari et al., 2013).

### 6.2. Lentil (*Lens culinaris*)

Lentils have been incorporated into different cuisines worldwide, especially in the Mediterranean and Indian regions (Faris et al., 2013). Likewise, lentils were mentioned in ancient treatment remedies; in traditional Indian medicine, they were used as a blood purifier, diuretic, and antifungal. Lentil is one of the herbs recommended in traditional Iranian medicine to treat menorrhagia (Shafiee & Nojavan, 2018). In the popular medicine of many ethnic groups, lentils are used to treat different diseases. They are used orally to treat diabetes and topically as a watery paste to treat skin infections and burns after roasting, grinding, and applying it directly to the affected areas. Lentils also treat various kidney and gastric ailments (Adikay & Saisruthi, 2016; Faris et al., 2013).

### 6.3. Lupine (*Lupinus termis*)

Lupine is a crop that can grow on marginal agricultural land in various environments and conditions, so it has a high potential to contribute significantly to existing food sources; this crop has medicinal characteristics (Khan et al., 2015). In Unani medicine, the seeds are considered anti-inflammatory, diuretic, emmenagogue, and anthelmintic; these are also applied externally to resolve inflammations. Lupine seeds are used in traditional medicine in Africa and the Middle East as an antidiabetic agent and topically to treat acne. Also, it is traditionally used in the USA for eczema of the feet (Alarjani et al., 2022).

### 6.4. Pea (*Pisum sativum*)

Peas are considered one of the most important leguminous crops as they are cultivated in 84 different countries. Dried beans are consumed in various forms, such as soup, chat, cola, dal, stew, snacks, vegetables, and flour. Dried pea flour is valued not only as an important source of dietary protein but also for its industrial use as a thickening agent in certain food industries. In contrast, green seeds are used as fresh, frozen, or canned vegetables (Parihar et al., 2016). In addition, the pea has been used in traditional medicine. In ancient traditional Chinese medicine, the pea was recorded to have effects in treating diabetes (Lu et al., 2019; Qin et al., 2021). Also, traditionally, the seeds are used as a laxative, astringent, contraceptive, fungistatic, and spermicide to treat wrinkled skin, diabetes, acne, phlegm, intestinal inflammation, diuretic, vomiting, hiccups, belching, constipation, and cough. The pea also decreases the incidence of cancer and cardiovascular diseases, reduces the chances of skin eruptions, and accelerates healing. Finally, the pod is used for fungal infections, and the seed oil, given once a month to women, shows promise in preventing pregnancy (Rehman & Khanum, 2011; Dazal et al., 2015; Zilani et al. al, 2017).

**Table 1.** Therapeutic uses of legumes in traditional medicine

| Legume | Used Part | Therapeutic Uses | Ref |
|---|---|---|---|
| Common Bean | Seeds | Pharyngitis, Rapid umbilical cord removal in newborns, Antidiabetic | Loko et al., 2018; Helmstädter, 2010 |
| Common beans | Leaves | Babies who cannot breastfeed, wounds, vaginal infections, bad body odor, bee stings, sterility, fever | |
| Common beans | Roots | Obesity, difficult childbirth | |

| Legume | Used Part | Therapeutic Uses | Ref |
|---|---|---|---|
| Soybean | Seeds | Improves heart health, hypocholesterolemic, improves digestive tract function, prevents cancer (prostate, breast, colon), bone health, immunostimulator, antiatherogenic, antioxidant, regulates lipid metabolism, | Ahmad et al., 2014; Sugano, 2006; Nyandwaro et al., 2020; Jung et al., 2020 |
| Chickpea | Seeds | Body weakness, lung enlargement, lung ulcers, antiparasitic, aphrodisiac, impotence, abortifacient, antidote, kidney stones, diuretic, cancer, breast milk production, diabetes, hyperlipidemia, hypertension, bronchitis, coprosis, loss of appetite, itching skin, flatulence, osteoporosis, tumor formation, obesity, blood dyscrasias, ear infections, weight gain, anti-dandruff. | Ahmadi et al., 2020; Aisa et al., 2019; Heydarirad et al., 2019; Kumar et al., 2022; Memariani et al., 2020; Rachwat-Rosiak et al., 2015 |
| Chickpea | Seed oil | Joint pain | Abbasifard et al., 2020 |
| Chickpea | Leaves | Menstrual and urinary disorders, warts, jaundice, gout pain, indigestion, constipation, dislocations. | Gupta et al., 2016; Kumar et al., 2022 |
| Pigeon pea | Seeds | Analgesic, sedative, and antivertiginous help with urinary incontinence in men and treating kidney diseases. | Tiwari et al., 2013 |
| Pigeon pea | Leaves | Stop bleeding, analgesic, antihelmintic, food poisoning, constipation, burns, dizziness and diabetes, anemia, hepatitis, measles, chickenpox, gonorrhea, diarrhea, epilepsy, and stabilize the menstrual period, in ulcers and oral inflammations, in sores, wounds, and abdominal tumors. | Cui et al., 2015; Dinore & Farooqui, 2020; Pal et al., 2011; Srivastava, 2013 |
| Pigeon pea | Stems | Gingivitis, stomatitis, wounds and analgesic. | Pal et al., 2011 |
| Pigeon pea | Roots | Cough, stomach problems, syphilis, and analgesics. | Srivastava, 2013 |
| Lentil | Seeds | Blood purifier, diuretic, antifungal, kidney and gastric ailments, and menorrhagia. | Adikay & Saisruthi, 2016; Shafiee & Nojavan, 2018 |
| Lupine | Seeds | Anti-inflammatory, diuretic, emmenagogue, anthelmintic, antidiabetic, and used topically to treat acne and eczema of the feet. | Alarjani et al., 2022 |
| Pea | Seeds | Diuretics are mild laxatives for vomiting, hiccups, belching, constipation, and cough, reduce the chances of skin rashes, accelerate healing, and treat diabetes. | Ahmed & Hasan, 2014; Qin et al., 2021 |

## Conclusion

The medicinal properties of legumes have been proven in traditional medicine for many years, and this has been part of the development and application of modern medicine. Nevertheless, further research on the therapeutic benefits of each legume is necessary, including even those that are not commonly consumed and found only in some regions of the world. It can be concluded that legumes are functional foods that contain macronutrients and several compounds with biological activity that can be used in various diseases. However, the knowledge of traditional medicine is also part of the growth of modern medicine.

## References

Abbasifard, M., Bazzaz, A., Bazmandegan, G., Rezaeian, M., Saeedaskari, P., Mahmoodian, H., & Zareshahi, R. (2020). Effect of topical chickpea oil (*Cicer arietinum* L.) on knee osteoarthritis: A randomized double-blind controlled clinical trial. *European Journal of Integrative Medicine*, *35*, 101076. doi: 10.1016/j.eujim.2020.101076.

Adikay, S. (2016). Phytoremedial effect of *Lens culinaris* against doxorubicin-induced nephrotoxicity in male Wistar rats. *International Journal of Green Pharmacy (IJGP)*, *10*(03). doi: 10.22377/ijgp.v10i03.678.

Ahmad, A., Hayat, I., Arif, S., Masud, T., Khalid, N., & Ahmed, A. (2014). Mechanisms Involved in the Therapeutic Effects of Soybean (*Glycine max*). *International Journal of Food Properties*, *17*(6), 1332-1354. doi:10.1080/10942912.2012.714828.

Ahmadi, N., Mokaberinejad, R., Saeidi, A., Zandi, A., Leach, M. J., & Pasalar, M. (2020). The effect of chickpea broth on knee osteoarthritis—A Pilot non-randomised open-labeled clinical study. *Advances in Integrative Medicine*, *7*(3), 121-125. doi:10.1016/j.aimed.2020.01.004.

Ahmed, S., & Hasan, M. M. (2014). Legumes: an overview. *RADS Journal of Pharmacy and Pharmaceutical Sciences*, *2*(1), 34-38.

Aisa, H. A., Gao, Y., Yili, A., Ma, Q., & Cheng, Z. (2019). Chapter 39 - Beneficial Role of Chickpea (*Cicer arietinum* L.) Functional Factors in the Intervention of Metabolic Syndrome and Diabetes Mellitus. In R. R. Watson & V. R. Preedy (Eds.), *Bioactive Food as Dietary Interventions for Diabetes (Second Edition)* (pp. 615-627). Academic Press. doi:10.1016/B978-0-12-813822-9.00039-4.

Alarjani, K. M., Huessien, D., Rasheed, R. A., & Kalaiyarasi, M. (2022). Green synthesis of silver nanoparticles by *Pisum sativum* L. (pea) pod against multidrug resistant foodborne pathogens. *Journal of King Saud University - Science*, *34*(3), 101897. doi:10.1016/j.jksus.2022.101897.

Ali, S. E., El Badawy, S. A., Elmosalamy, S. H., Emam, S. R., Azouz, A. A., Galal, M. K., Abd-Elsalam, R. M., Issa, M. Y., & Hassan, B. B. (2021). Novel promising reproductive and metabolic effects of *Cicer arietinum* L. extract on letrozole induced polycystic ovary syndrome in rat model. *Journal of Ethnopharmacology*, *278*, 114318. doi:10.1016/j.jep.2021.114318.

Anderson, E. J., Ali, M. L., Beavis, W. D., Chen, P., Clemente, T. E., Diers, B. W., Graef, G. L., Grassini, P., Hyten, D. L., McHale, L. K., Nelson, R. L., Parrott, W. A., Patil, G. B., Stupar, R. M., & Tilmon, K. J. (2019). Soybean [*Glycine max* (L.) Merr.] Breeding: History, Improvement, Production and Future Opportunities. In J. M. Al-Khayri, S. M. Jain, & D. V. Johnson (Eds.), *Advances in Plant Breeding Strategies: Legumes: Volume 7* (pp. 431-516). Springer International Publishing. doi:10.1007/978-3-030-23400-3_12.

Bitocchi, E., Bellucci, E., Giardini, A., Rau, D., Rodriguez, M., Biagetti, E., Santilocchi, R., Spagnoletti Zeuli, P., Gioia, T., Logozzo, G., Attene, G., Nanni, L., & Papa, R. (2013). Molecular analysis of the parallel domestication of the common bean (*Phaseolus vulgaris*) in Mesoamerica and the Andes. *New Phytologist*, *197*(1), 300-313. doi:10.1111/j.1469-8137.2012.04377.x.

Bljajić, K., Petlevski, R., Vujić, L., Čačić, A., Šoštarić, N., Jablan, J., Saraiva de Carvalho, I., & Zovko Končić, M. (2017). Chemical Composition, Antioxidant and α-Glucosidase-Inhibiting Activities of the Aqueous and Hydroethanolic Extracts of *Vaccinium myrtillus* Leaves. *Molecules*, *22*(5). doi:10.3390/molecules22050703.

Borges, L., Bruneau, A., Cardoso, D., Crisp, M., Delgado-Salinas, A., Doyle, J. J., Egan, A., Herendeen, P. S., Hughes, C., Kenicer, G., Klitgaard, B., Koenen, E., Lavin, M., Lewis, G., Luckow, M., Mackinder, B., Malécot, V., Miller, J. T., Pennington, R. T., . . . Wink, M. (2013). Towards a new classification system for legumes: Progress report from the 6th International Legume Conference. *South African Journal of Botany*, *89*, 3-9. doi:10.1016/j.sajb.2013.07.022.

Craig, O. E., Steele, V. J., Fischer, A., Hartz, S., Andersen, S. H., Donohoe, P., Glykou, A., Saul, H., Jones, D. M., Koch, E., & Heron, C. P. (2011). Ancient lipids reveal continuity in culinary practices across the transition to agriculture in Northern Europe. *Proceedings of the National Academy of Sciences of the United States of America*, *108*(44), 17910-17915. doi:10.1073/pnas.1107202108.

Cristiani, E., Radini, A., Edinborough, M., & Borić, D. (2016). Dental calculus reveals Mesolithic foragers in the Balkans consumed domesticated plant foods. *Proceedings of the National Academy of Sciences*, *113*(37), 10298-10303. doi:10.1073/pnas.1603477113.

Cui, Q., Peng, X., Yao, X.-H., Wei, Z.-F., Luo, M., Wang, W., Zhao, C.-J., Fu, Y.-J., & Zu, Y.-G. (2015). Deep eutectic solvent-based microwave-assisted extraction of genistin, genistein, and apigenin from pigeon pea roots. *Separation and Purification Technology*, *150*, 63-72. doi:10.1016/j.seppur.2015.06.026.

Damián-Medina, K., Salinas-Moreno, Y., Milenkovic, D., Figueroa-Yáñez, L., Marino-Marmolejo, E., Higuera-Ciapara, I., Vallejo-Cardona, A., & Lugo-Cervantes, E. (2020). In silico analysis of antidiabetic potential of phenolic compounds from blue corn (*Zea mays* L.) and black bean (*Phaseolus vulgaris* L.). *Heliyon*, *6*(3), e03632. doi:10.1016/j.heliyon.2020.e03632.

Dazal, M., Gonzales, E., Sabino, A., Salazar, R., Samar, B., San Juan, M., & Castillo, A. (2015). Controlling biofilm formation by inhibiting the quorum-sensing activity of *Pseudomonas aeruginosa* using the ethanolic extracts of *Piper nigrum* (Piperaceae) fruit, *Punica granatum* (Lythraceae) Pericarp, and *Pisum sativum* (Fabaceae) seed. *International Journal of Pharmaceutical Sciences and Drug Research*, *7*, 349-353.

Didinger, C., & Thompson, H. J. (2021). Defining Nutritional and Functional Niches of Legumes: A Call for Clarity to Distinguish a Future Role for Pulses in the Dietary Guidelines for Americans. *Nutrients*, *13*(4). doi:10.3390/nu13041100.

Dinore, J. M., & Farooqui, M. (2022). GC-MS and LC-MS: an integrated approach towards the phytochemical evaluation of methanolic extract of Pigeon Pea [*Cajanus cajan (L.) Millsp*] leaves. *Nat Prod Res*, *36*(8), 2177-2181. doi:10.1080/14786419.2020.1849197.

Duranti, M. (2006). Grain legume proteins and nutraceutical properties. *Fitoterapia*, *77*(2), 67-82. doi:10.1016/j.fitote.2005.11.008.

FAO. (2023). *Soybean*. Retrieved July 7th from https://www.fao.org/land-water/databases-and-software/crop-information/soybean/en/.

Faris, M. e. A.-I. E., Takruri, H. R., & Issa, A. Y. (2013). Role of lentils (*Lens culinaris* L.) in human health and nutrition: a review. *Mediterranean Journal of Nutrition and Metabolism*, *6*(1), 3-16. doi:10.1007/s12349-012-0109-8.

Giangrieco, I., Tamburrini, M., Tuppo, L., Pasquariello, M. S., & Ciardiello, M. A. (2022). Healthy biological activities in legume flours from industrial cooking. *Food Bioscience*, *48*, 101743. doi:10.1016/j.fbio.2022.101743.

Graham, P. H., & Vance, C. P. (2003). Legumes: Importance and Constraints to Greater Use. *Plant Physiology*, *131*(3), 872-877. doi:10.1104/pp.017004.

Gupta, R. K., Gupta, K., Sharma, A., Das, M., Ansari, I. A., & Dwivedi, P. D. (2016). Health risks and benefits of chickpea (*Cicer arietinum*) consumption. *Journal of Agricultural and Food Chemistry*, *65*(1), 6-22. doi:10.1021/acs.jafc.6b02629.

Güzel, D., & Sayar, S. (2012). Effect of cooking methods on selected physicochemical and nutritional properties of barlotto bean, chickpea, faba bean, and white kidney bean. *Journal of Food Science and Technology*, *49*(1), 89-95. doi:10.1007/s13197-011-0260-0.

Hayat, I., Ahmad, A., Masud, T., Ahmed, A., & Bashir, S. (2014). Nutritional and health perspectives of beans (*Phaseolus vulgaris* L.): an overview. *Critical reviews in food science and nutrition*, *54*(5), 580-592. doi:10.1080/10408398.2011.596639.

Helmstädter, A. (2010). Beans and diabetes: *Phaseolus vulgaris* preparations as antihyperglycemic agents. *Journal of Medicinal Food*, *13*(2), 251-254. doi:10.1089/jmf.2009.0002.

Henry, A. G., Brooks, A. S., & Piperno, D. R. (2011). Microfossils in calculus demonstrate consumption of plants and cooked foods in Neanderthal diets (Shanidar III, Iraq; Spy I and II, Belgium). *Proceedings of the National Academy of Sciences*, *108*(2), 486-491. doi:10.1073/pnas.1016868108.

Heydarirad, G., Choopani, R., Pasalar, M., Parvizi, M. M., Hajian, P., & Mirzaei, H. R. (2019). The Effect of a Chickpea-Based Persian Diet on Cancer-Related Fatigue in Breast Cancer Patients: A Semi-Experimental Study. *Complementary Medicine Research*, *26*(6), 390-397. doi:10.1159/000500017.

Jung, Y. S., Rha, C.-S., Baik, M.-Y., Baek, N.-I., & Kim, D.-O. (2020). A brief history and spectroscopic analysis of soy isoflavones. *Food Science and Biotechnology*, *29*(12), 1605-1617. doi:10.1007/s10068-020-00815-6.

Katiyar, A., & Mishra, S. (2015). Nutritional and biochemical importance of chickpea in respect to human health: a review. *Journal of Plant Development Sciences*, *7*(5), 415-420.

Khan, M. K., Karnpanit, W., Nasar-Abbas, S. M., Huma, Z. e., & Jayasena, V. (2015). Phytochemical composition and bioactivities of lupin: a review. *International Journal of Food Science & Technology*, *50*(9), 2004-2012. doi:10.1111/ijfs.12796.

Kizilarslan-Hançer, C., Sevgi, E., Akkaya, M., & Altundağ-Çakir, E. (2019). As a living culture of traditional herbal coffee in Turkey: Chickpea coffee. *Düzce Üniversitesi Bilim ve Teknoloji Dergisi*, *7*(1), 239-247. doi:10.29130/dubited.435752.

Kumar, R., Singh, R. K., Misra, J., Yadav, A., Kumar, A., Yadav, R., Hegde, V., Kumar, S., & Yadav, N. (2022). Dissecting proteomic estimates for enhanced bioavailable nutrition during varied stages of germination and identification of potential genotypes in chickpeas. *Legume Research-An International Journal*, *45*(9), 1082-1087. doi:10.18805/LR-4531.

Lee, K. J., Baek, D.-Y., Lee, G.-A., Cho, G.-T., So, Y.-S., Lee, J.-R., Ma, K.-H., Chung, J.-W., & Hyun, D. Y. (2020). Phytochemicals and Antioxidant Activity of Korean Black Soybean (*Glycine max* L.) Landraces. *Antioxidants*, *9*(3), 213 doi: 10.3390/antiox9030213.

Li, P., Lu, B., Gong, J., Li, L., Chen, G., Zhang, J., Chen, Y., Tian, X., Han, B., Guo, Y., Xie, Z., & Liao, Q. (2021). Chickpea Extract Ameliorates Metabolic Syndrome Symptoms via Restoring Intestinal Ecology and Metabolic Profile in Type 2 Diabetic Rats. *Molecular Nutrition and Food Research*, *65*(13), e2100007. doi:10.1002/mnfr.202100007.

Loko, L. E. Y., Toffa, J., Adjatin, A., Akpo, A. J., Orobiyi, A., & Dansi, A. (2018). Folk taxonomy and traditional uses of common bean (*Phaseolus vulgaris* L.) landraces by the sociolinguistic groups in the central region of the Republic of Benin. *Journal of Ethnobiology and Ethnomedicine*, *14*(1), 52. doi:10.1186/s13002-018-0251-6.

Lomas-Soria, C., Pérez-Ramírez, I. F., Caballero-Pérez, J., Guevara-Gonzalez, R. G., Guevara-Olvera, L., Loarca-Piña, G., Guzman-Maldonado, H. S., & Reynoso-Camacho, R. (2015). Cooked common beans (*Phaseolus vulgaris* L.) modulate renal genes in streptozotocin-induced diabetic rats. *Journal of Nutritional Biochemistry*, *26*(7), 761-768. doi:10.1016/j.jnutbio.2015.02.006.

Lu, Z. X., He, J. F., Zhang, Y. C., & Bing, D. J. (2020). Composition, physicochemical properties of pea protein and its application in functional foods. *Critical Reviews in Food Science and Nutrition*, *60*(15), 2593-2605. doi:10.1080/10408398.2019.1651248.

Memariani, Z., Gorji, N., Moeini, R., & Farzaei, M. H. (2020). Traditional uses. In *Phytonutrients in Food* (pp. 23-66). Elsevier. doi:10.1016/B978-0-12-815354-3.00004-6.

Nyandwaro, K., Oyweri, J., Kimani, F., & Mbugua, A. (2020). Evaluating Antiplasmodial and Antimalarial Activities of Soybean (*Glycine max*) Seed Extracts on *P. falciparum*

Parasite Cultures and *P. berghei* Infected Mice. *Journal of Pathogens, 2020*, 7605730. doi:10.1155/2020/7605730.

Ogu, J. M. (2022). Ethics of folk medicine among the Igbo. *Developing World Bioethics, 22*(4), 203-210. doi:10.1111/dewb.12330.

Pal, D., Mishra, P., Sachan, N., & Ghosh, A. K. (2011). Biological activities and medicinal properties of *Cajanus cajan* (L) Millsp. *Journal of Advanced Pharmaceutical Technology & Research, 2*(4), 207-214. doi:10.4103/2231-4040.90874.

Parihar, A. K., Bohra, A., & Dixit, G. P. (2016). Nutritional Benefits of Winter Pulses with Special Emphasis on Peas and Rajmash. In U. Singh, C. S. Praharaj, S. S. Singh, & N. P. Singh (Eds.), *Biofortification of Food Crops* (pp. 61-71). Springer India. doi:10.1007/978-81-322-2716-8_6.

Petran, M., Dragos, D., & Gilca, M. (2020). Historical ethnobotanical review of medicinal plants used to treat children diseases in Romania (1860s-1970s). *Journal of Ethnobiology and Ethnomedicine, 16*(1), 15. doi:10.1186/s13002-020-00364-6.

Phumthum, M., Balslev, H., Kantasrila, R., Kaewsangsai, S., & Inta, A. (2020). Ethnomedicinal Plant Knowledge of the Karen in Thailand. *Plants (Basel), 9*(7). doi:10.3390/plants9070813.

Polito, L., Bortolotti, M., Maiello, S., Battelli, M. G., & Bolognesi, A. (2016). Plants Producing Ribosome-Inactivating Proteins in Traditional Medicine. *Molecules, 21*(11). doi:10.3390/molecules21111560.

Przeor, M. (2022). Some Common Medicinal Plants with Antidiabetic Activity, Known and Available in Europe (A Mini-Review). *Pharmaceuticals, 15*(1).

Qin, G., Xu, W., Liu, J., Zhao, L., & Chen, G. (2021). Purification, characterization and hypoglycemic activity of glycoproteins obtained from pea (*Pisum sativum* L.). *Food Science and Human Wellness, 10*(3), 297-307. doi:10.1016/j.fshw.2021.02.021.

Rachwa-Rosiak, D., Nebesny, E., & Budryn, G. (2015). Chickpeas—composition, nutritional value, health benefits, application to bread and snacks: a review. *Critical reviews in food science and nutrition, 55*(8), 1137-1145. doi:10.1080/10408398.2012.687418.

Rehman, S., & Khanum, A. (2011). Isolation and characterization of peptide (s) from *Pisum sativum* having antimicrobial activity against various bacteria. *Pakistan Journal of Botany, 43*, 2971-2978.

Sahu, M., Vermaand, D., & Harris, K. (2014). Phytochemical analysis of the leaf, stem and seed extracts of *Cajanus cajan* L (Dicotyledoneae: Fabaceae). *World Journal of Pharmacy and Pharmaceutical Sciences, 3*(8), 694-733.

Setchell, K. D. (2017). The history and basic science development of soy isoflavones. *Menopause, 24*(12), 1338-1350. doi: 10.1097/GME.0000000000001018.

Shafiei, M. (2018). Therapeutic effects of roasted lentil flour from the viewpoint of Iranian Islamic resources: a review. *Health, Spirituality and Medical Ethics, 5*(4), 53-58. doi:10.29252/jhsme.5.4.53.

Simonsen, A. K., Dinnage, R., Barrett, L. G., Prober, S. M., & Thrall, P. H. (2017). Symbiosis limits the establishment of legumes outside their native range at a global scale. *Nature Communications, 8*(1), 14790. doi:10.1038/ncomms14790.

Srivastava, J. (2013). In-vitro regeneration protocol for pigeon pea—A review. *International Journal of Food, Agriculture and Veterinary Sciences*, *3*(2), 63. doi: 10.1007/s12298-011-0079-1.

Sugano, M. (2006). *Soy in Health and Disease Prevention*. CRC Press. doi:10.1201/9781420026566.

Tiwari, A. K., Abhinay, B., Babu, K. S., Kumar, D. A., Zehra, A., & Madhusudana, K. (2013). Pigeon pea seed husks as potent natural resource of anti-oxidant and anti-hyperglycaemic activity. *International Journal of Green Pharmacy (IJGP)*, *7*(3). doi:10.4103/0973-8258.120247.

Tran, N., Pham, B., & Le, L. (2020). Bioactive Compounds in Anti-Diabetic Plants: From Herbal Medicine to Modern Drug Discovery. *Biology*, *9*(9), 252 doi: 10.3390/biology9090252.

Tungmunnithum, D., & Hano, C. (2020). The cosmetic potential of *Cajanus cajan* (L.) millsp: Botanical data, traditional uses, phytochemistry, and biological activities. *Cosmetics*, *7*(4), 84. doi:10.3390/cosmetics7040084.

Wei, Y., Li, P., Li, B., Gao, J., Wang, D., Qin, L., Sun, W., Xu, Y., Shi, H., Xu, T., & Liu, T. (2017). Study of the Hypoglycemic Activity of Derivatives of Isoflavones from *Cicer arietinum* L. *Evidence-Based Complementary and Alternative Medicine*, *2017*, 8746823. doi:10.1155/2017/8746823.

WHO, W. H. O. (2023). *Traditional, Complementary and Integrative Medicine*. Retrieved 09 June, 2023 from https://www.who.int/health-topics/traditional-complementary-and-integrative-medicine#tab=tab_1.

Zilani, M. N., Sultana, T., Asabur Rahman, S. M., Anisuzzman, M., Islam, M. A., Shilpi, J. A., & Hossain, M. G. (2017). Chemical composition and pharmacological activities of *Pisum sativum*. *BMC Complement Altern Med*, *17*(1), 171. doi:10.1186/s12906-017-1699-y.

# Chapter 2

# Common Beans (*Phaseolus vulgaris*): Composition and Functional Applications

**M. Perucini-Avendaño**[1]
**E. P. Díaz-Galindo**[2]
**M. Sánchez-Becerril**[3]
**B. H. Camacho-Díaz**[4]
**and M. Nicolás-García**[5,*]
[1]Programa de Investigación y Posgrado en Ciencia de los Alimentos, Facultad de Química, Universidad Autónoma de Querétaro, Santiago de Querétaro, Querétaro, México
[2]Departamento de Ingeniería Bioquímica, Escuela Nacional de Ciencias Biológicas, Instituto Politécnico Nacional (IPN), Unidad Profesional Adolfo López Mateos, Zacatenco, Delegación Gustavo A. Madero, Ciudad de México
[3]Tecnológico Nacional de México/Tecnológico de Estudios Superiores de San Felipe del Progreso, División de Ingeniería en Industrias Alimentarias, San Felipe del Progreso, Estado de México, México
[4]CEPROBI-Instituto Politécnico Nacional, Carretera Yautepec-Jojutla, Colonia San Isidro, Yautepec, Morelos, Mexico
[5]Tecnológico Nacional de México/Instituto Tecnológico Superior de Teziutlán, Ingeniería en Industrias Alimentarias, Teziutlán, Puebla, Mexico

## Abstract

The *Fabaceae* family has more than 20,000 species and more than 700 genera, including the genus *Phaseolus*. The common bean (*Phaseolus vulgaris* L.) has been one of the most emblematic foods in Mexican

---

* Corresponding Author's Email: mnicolasg25@hotmail.com

In: Legumes: Beyond Their Nutritional Value
Editors: C. Jiménez-Martínez, M. F. Juárez-Chairez and M. S. Cid-Gallegos
ISBN: 979-8-89530-440-2

culture since pre-Hispanic times due to its nutritional attributes as an excellent source of carbohydrates, protein, fiber, minerals, and phytochemicals. It contains bioactive compounds, such as phenolic compounds, peptides, and anthocyanins, that have a significant metabolic role and beneficial effects on the organisms of people consuming this food. All commercial bean varieties undergo various processing methods prior to consumption. These processes can significantly influence the bioactive compounds within the seeds, thereby altering their biological activity. Over time, a growing body of scientific evidence, supported by numerous experimental models, has emerged to evaluate and isolate these bioactive compounds, which have shown potential in addressing chronic degenerative diseases, including cardiovascular conditions, weight management, and stress reduction. This chapter aims to elucidate the chemical composition of beans—both nutritional and non-nutritional—their impact on health, and the processing technologies employed in bean preparation, with a particular focus on how these technologies affect the retention and alteration of bioactive compounds.

**Keywords**: common bean, bioactive compounds, extraction methods

## 1. Introduction: Common Bean Origin and Production

*Phaseolus vulgaris L.*, traditionally called bean, is a plant from the Fabaceae family, *Papilionoideae* subfamily, tribe *Phaseoleae*, subtribe *Phaseolinae,* and comprises about 70 species (Rodríguez et al., 2022). These plants were cultivated over 8000 years ago (Ganesan and Xu, 2017). Mesoamerica has been indicated as the origin and domestication center of beans, which migrated from this area toward South America. Mexico is considered the origin of 47 out of the 52 recognized species (Chávez-Mendoza and Sánchez, 2017).

Based on their morphological characteristics, beans are categorized under various names, such as navy beans, kidney beans, red beans, black beans, pinto beans, and cranberry beans (Ganesan and Xu, 2017). Additionally, each country assigns specific local names to the beans cultivated in its region. The primary producers of beans are located in Latin America, the Caribbean, Africa, and Asia, accounting for an estimated 77% of global production. Bean crops thrive across a wide geographical range, from 52° North latitude to 32° South latitude, at altitudes from sea level to over 3000 meters, and in climates that are largely free from the risk of frost (FAO, 2018).). The crop growing area is about 26.5 million hectares with an average yield of 697 kg per hectare; world climate change has affected the annual rainfalls, and environmental stresses such as drought, salinity, pests, and diseases harm global profits (Kulaz et al., 2023).

Beans are annual plants commonly cultivated in temperate and semi-tropical regions (Venegas et al., 2021). Environmental and soil conditions play a crucial role in determining the varieties of beans, which exhibit diversity in seed size, shape, and color (Shamseldin and Velázquez, 2020). Furthermore, the bean's ability to adapt to a wide range of environments has led to the development of new landraces, each with distinct structural variability, yield potential, stability, and nutritional value. Agronomic practices have also contributed to the vast biodiversity of bean varieties, characterized by different patterns of disease resistance and stress tolerance (Bosmali et al., 2023). The plant's green leaves, pods, and both immature and dry seeds are used as food. Notably, dry seeds are particularly valued for their ease of long-term storage, high nutritional value, and affordability (Mikić, 2018).

All bean plant parts are consumed, and even the straws are utilized for livestock feed (Venegas et al., 2021). It is the most consumed legume, with an estimated 30 million tons per year (Recek et al., 2021), consumed by 300 million people around the world (Lugo-Cervantes and Mojica, 2020), and it constitutes the main protein source in developing countries of Latin América and Africa (Bosmali et al., 2023). Beans are consumed in soups and salads or combined with other ingredients; they play an important role in the ready-to-eat food industry and meat-type derived products; in addition, beans show potential applications in food since their starches and proteins present good solubility, water/oil holding capacity, viscosity, and emulsion ability (Cominelli et al., 2022).

## 2. Nutritional Composition

Chemical composition, physical characteristics, and nutritional and non-nutritional content are closely related to seeds' genetic type, storage, and processing conditions (Bosmali et al., 2023; Corzo-Ríos et al., 2020). The general composition of the beans is from highest to lowest: carbohydrates, proteins, lipids, and minerals (Table 1). Bean proteins have a digestibility rate of approximately 79%, while their carbohydrates are characterized by slow digestion and a significant content of non-digestible carbohydrates (Chávez-Mendoza and Sánchez, 2017). Beans contain antioxidants, vitamins, carotenoids, phenolic compounds (high concentration in dark beans compared with light varieties), flavonoids, amino acids, sucrose, citric acid, and malic acid (Bosmali et al., 2023; Hernández-Guerrero et al., 2021; Lugo-cervantes and Mojica, 2020; Santos and Furlong, 2022).

## 2.1. Non-Nutritional Composition

Anti-nutritional compounds in plants serve as defense mechanisms against both biotic factors, such as bacteria, fungi, and pests, as well as abiotic factors, including UV light and adverse environmental conditions (Cominelli et al., 2022). Trypsin inhibitors, lectins, saponins, phytic acid, tannins, and oligosaccharides could show adverse nutritional and health effects when ingested at determined amounts and are associated with low protein and carbohydrate digestibility rate, toxicity potential at elevated concentrations, and pharmacological activities (Corzo-Ríos et al., 2020).

**Table 1.** General bean composition and main non-nutritional components

| Nutritional Component | |
|---|---|
| Carbohydrates | Starch is a major fraction in the bean seed (more than 50%); fiber is constituted by mono-, di-, and oligosaccharides (23.32-46.77%); the main components are pectins, pentosans, hemicellulose, cellulose, and lignin. |
| Proteins | Lysine (6.4-7.6 g/100 g protein), Phenylalanine plus tyrosine (5.3-8.2 g/100 g protein). |
| Lipid | A mixture of mono and polyunsaturated acylglycerides |
| Minerals | Iron, zinc |
| Non-Nutritional Component | |
| Trypsin inhibitors | Major non-nutritional constituent, protein<br>Intake reduction by inhibiting proteinase enzymes, low-rate food absorption, and acting as allergens, act as a protective agent in the dry seed. |
| Tannins | Low-rate food absorption and growth affect the bioavailability of minerals. |
| Lectins | Induce pancreas growth generating ulceration and intestinal epithelium necrosis in mice. |
| Phytic acid | Low bioavailability of minerals, proteins, and amino acids. |
| Oligosaccharides | Metabolized by bacteria producing methane, hydrogen, and carbon dioxide, and causing flatulence and diarrhea. |
| Polyphenols | Interference in nutrient absorption. |
| Saponins | Gastrointestinal complications. |
| Bioactive peptides and enzyme inhibitors | Interference in protein digestibility and bioavailability. |
| α-amylase inhibitor | Starch absorption reduction in livestock feeding. |

Source: Chávez-Mendoza and Sánchez, 2017; Corzo-Ríos et al. 2020; Lugo-cervantes and Mojica, 2020; Bogweh and Ageyo, 2021; Cominelli et al. 2022; Bosmali et al. 2023.

The main non-nutritional components and their characteristics are summarized in Table 1.

Nevertheless, these compounds also exhibit beneficial properties, including chemopreventive, antiproliferative, antimutagenic, anticarcinogenic, and antiapoptotic actions. Additionally, they possess antioxidant, hypocholesterolemic, hypoglycemic, antihypertensive, and antithrombotic properties, which are associated with reducing the risk of chronic diseases and enhancing endothelial function (Bosmali et al., 2023; Corzo-Ríos et al., 2020). Due to their high content of prebiotics and nutraceuticals, beans are considered a functional food (Hernández-Guerrero et al., 2021). Cooking at high temperatures is the general method for processing beans and reducing their phytic acid, oligosaccharides, minerals, tannins, and phenols content; on the other hand, this method notoriously improves the protein quality and soluble dietary fiber (SDF) content, decreasing the insoluble dietary fiber (IDF) content (Duenas et al., 2016).

## 3. Bioactive Compounds and Functional Properties

The common bean is a source of diverse compounds with biological activity, such as proteins, protease inhibitors, lipids, complex carbohydrates, dietary fiber, vitamins, resistant starch, oligosaccharides, α-galactosides, minerals, phytoestrogens, alkaloids, phenolic compounds, phytates, and saponins. These bioactive compounds have biological properties such as antioxidant, anticarcinogenic, antihypertensive, and prevention of non-communicable, chronic degenerative, cardiovascular, and osteoporosis diseases. In addition, emulsifying capacity measures the surface activity of carbohydrates and proteins to reduce the interfacial tension between two immiscible phases. In proteins, this activity depends on the presence of hydrophobic amino acids that enhance the stability of the emulsion to resist long-term interfacial changes. However, these properties are influenced by the processing method and conditions like temperature, pH, and sample ratio (Al-Ruwaih et al., 2019; Pedrosa et al., 2020; Lima-Becerra et al., 2021; Vogelsang-O'Dwyer et al., 2022; Ge et al., 2023). The functional properties of common bean-based food products are shown in Table 2.

Lima-Becerra et al. (2021) evaluated the foaming behavior and emulsifying capacity of beans based on different processing methods. Their findings revealed no significant differences in nutritional properties between traditional cooking and Ohmic heating. However, Ohmic heating was found to enhance the bean's techno-functional properties and reduce the content of trypsin inhibitors. Alfaro-Díaz et al. (2021) observed that high-temperature and pressure cooking (HTPC) (100°C for 15 minutes; 121°C for 30 minutes) and the extrusion process (90°C) of black beans increased protein permeability by 29–39%, along with improved techno-functional properties, particularly the emulsification capacity of protein concentrates. The emulsification capacity ranged from 556 to 971 g/g, depending on the pH (3.0–7.0), with pH 5 being the most favorable for enhancing emulsification (9.5–24.5%) in HTPC-treated samples. In contrast, extrusion-treated samples exhibited a 13.6% improvement in emulsification capacity at pH 7, compared to the untreated samples.

Additionally, Aguiló-Aguayo et al. (2021) evaluated the antioxidant and functional properties of protein isolates from the common bean variety Ganxet. The protein isolate, with a purity of 91%, demonstrated oil retention capacity of 2.76 g/g and water retention capacity of 1.25 g/g. The highest levels of techno-functional properties were observed at pH 8.0 and 10.0, where foaming capacities reached 86.25% and 78.75%, respectively. Furthermore, emulsion stability was reported at 94% and 93.9%, highlighting the great potential of Ganxet beans as valuable techno-functional ingredients for the development of novel food products.

### 3.1. Proteins: Protein Hydrolysates and Bioactive Peptides

Legumes such as beans are becoming an excellent source of protein, considered as an alternative functional ingredient of vegetable origin compared to commercial ingredients due to their physical and techno-functional properties, such as water solubility, water and oil retention capacity, emulsifying, gelling, and foaming capacity (Pedrosa et al., 2020; Shevkani et al., 2022; Vogelsang-O'Dwyer et al., 2022; De Paiva Gouvêa et al., 2023). Ashraf et al. (2020) reported a % protein content of 88.39% in white kidney protein concentrates, within the recommended content of commercial protein isolates (90%). Likewise, protein isolates have been reported to be used as emulsifiers with high foam stability (De Paiva Gouvêa et al., 2023).

Pedrosa et al. (2020) compared the techno-functional properties of the white bean variety Almonga protein isolates and the pea. The bean isolate had a high content of total protein (75.42%) and soluble protein (44.15%), with a higher foaming capacity (47%), swelling (5.42%), emulsifying (26.85%), and emulsion stability (28.70%). The properties identified were higher than those of the pea, including a high content of trypsin inhibitors (21.95 TIU/mg). In turn, De Paiva Gouvêa et al. (2023) obtained a protein concentrate with a particle size of 9.82 μm with a high foaming capacity of 111%, higher than other legumes such as the broad bean (93%), with a stability of 60 min.

Protein composition and structure influence the functional properties of beans (Vogelsang-O'Dwyer et al., 2022; De Paiva Gouvêa et al., 2023). It has been reported that the amino acid profile of white kidney beans is predominantly composed of essential amino acids, including leucine, lysine, phenylalanine, and tyrosine, all of which align with the FAO/WHO recommended values (Ashraf et al., 2020). Bioactive peptides consisting of 2 to 20 amino acids are the result of gastrointestinal digestion (enzymatic hydrolysis) or due to processing (cooking, germination, and fermentation) of proteins present in food (Al-Ruwaih et al., 2019; Ashraf et al., 2020; Sanchez-Chino et al., 2022). Al-Ruwaih et al. (2019) evaluated high-pressure treatment (300-600 MPa/15 min) and enzymatic hydrolysis (Alcalase: 0.5-1%, I/O) for obtaining protein hydrolysates. They observed that the foaming capacity increased, highlighting the one treated at 300 MPa with a capacity of 90.27%. However, the foam stability was affected due to the protein hydrolyzate's inability to maintain the polypeptide chains' lower surfactant activity.

### 3.2. Carbohydrates: Dietary Fiber and Starch

The highest carbohydrate content is found in the bean cotyledon, which contains 50-60%, starch being the major component (27-60%) and other components or non-starch polysaccharides (cellulose, hemicellulose, arabinoxylan, and lignin), and oligosaccharides (stachyose, raffinose, and verbascose). Due to their smaller size, oligosaccharides ferment rapidly, and their prebiotic effect has been demonstrated (Alcázar-Valle et al., 2020; Wainainaina et al., 2021; Perera et al., 2023).

On the other hand, dietary fiber (soluble and insoluble) and resistant starch in beans play an important role in health, mainly for diabetics, due to their low glycemic index compared to other carbohydrate-rich foods. It has been shown that incorporating bean flour is an alternative to obtaining functional foods and is present in more than 50% of the marketed foods. The soluble fiber that includes galactomannan gums, inulins, pectins, mucilages, oligosaccharides, and β-glucans participate in glycemic control and cholesterol-lowering (Ramírez-Jiménez et al., 2018; Nartea et al., 2023; Perera et al., 2023). The bean seed coat is constituted of 79.82% dietary fiber divided into soluble (3-9%) and insoluble (63-86%) (Nartea et al., 2023).

Ramírez-Jiménez et al. (2018) developed various formulations of a snack bar by incorporating common bean flour from the Bayo Madero variety. They observed that the carbohydrate contents of the bars were lower than 5 g/100g, and a 60% increase in dietary fiber impacted a decrease in energy content (11-12% kcal). Shevkani et al. (2022) observed that bean flour showed resistant starch (32.0-50.5%), water absorption capacity (1.7-2.7 g/g), fat absorption capacity (1.4-1. 7 g/g), emulsifying activity index (50.1-70.1 $m^2/g$), foaming capacity (70.8-98.3%), foam stability (82.4-91.3%), and dry matter digestibility (14.6-47.2%). They showed that the reduction in starch and dry matter digestibility is due to the contribution of starch-lipid complexes, proteins, and non-starch carbohydrates. Likewise, the techno-functional properties evaluated are due to starch, protein, resistant starch, and ash content. On the other hand, a study evaluated the *in vitro* starch digestibility of different beans grown in Brazil and found diverse behaviors. White kidney beans showed the highest content of resistant starch (7.2%), and in terms of rapid resistant starch, red kidney beans were the lowest (89.7%) (Demiate et al., 2016). Furthermore, another investigation evaluated the swelling power and solubility of starches from kidney bean cultivars, and they reported 1.62-2.13 (g/g) and 2.11-2.285% at 60°C, respectively (Gani et al., 2012).

### 3.3. Lipids: Unsaturated Fatty Acids and Vitamins

The lipids present in beans contribute to their high nutritional value. Despite the low content of 1-3%, depending on the variety, beans contain unsaturated (65% to 87%), polyunsaturated (30.8 to 43.4%), and saturated (10-15%) fatty acids that include linolenic, linoleic, palmitic, palmitoleic, stearic, eicosatrienoic, and elaidic acids (Grela et al., 1995; David et al., 2019, Alcázar-Valle et al., 2021; Wainaina et al., 2021; Martínez-Ávila et al., 2022). Other components of a lipid nature are carotenoids and tocopherols with antioxidant properties. These act as precursors of vitamins A and E (α, β, δ, γ tocopherol, and tocotrienol) and are found mainly in the bean seed coat, with γ-tocopherol standing out (Nartea et al., 2023).

### 3.4. Phenolic Compounds

Beans are considered a good source of phenolic compounds with antioxidant properties, mostly present in the testa; they are characteristic of plants and are known as secondary metabolites (Yang et al., 2018). In recent years, the impact of phenolic compounds on human health has been evidenced. Phenolic compounds are distributed throughout the bean seed, mainly in the seed coat that has the highest concentration (64.13% - 93.08%), which includes flavonoids, flavones, flavanols, flavanones, flavanones, isoflavonoids, anthocyanins, chalcones, dihydrochalcones, condensed tannins, anthocyanins, and flavonols.

The cotyledon is a source of phenolic acids such as ferulic, sinapic, chlorogenic, and other hydroxycinnamic acids present in smaller amounts (Finetti et al., 2020; Nicolás-García et al., 2021; Nicolás-García et al., 2022; Stefanucci et al., 2022; Nartea et al., 2023). The main function of these compounds is to be used for plant growth and reproduction, as well as to serve as a defense mechanism against danger, or predators; they are a complex group of substances with a wide range of molecular weight, which are either free, bound to the cell wall or fiber and food proteins (Goñi et al., 2009). To obtain phenolic extracts, grinding, degreasing, solvent extraction, centrifugation, filtration, evaporation, and drying must be performed (Yang et al., 2018). Several studies have shown a correlation between bean consumption and a lower risk of cardiovascular disease, obesity, and diabetes. These health benefits from bean consumption have been partially attributed to polyphenols with antioxidant capacity (Xu & Chang, 2009).

## 4. Applications in the Food Industry

Beans are recognized as a functional ingredient that enhances the nutritional and functional quality of food products. They are of particular interest to the food industry due to their widespread availability, rich nutrient profile, and affordability (Ramírez-Jiménez et al., 2018; Pedrosa et al., 2020; Lima-Becerra et al., 2021). Carboni et al. (2023) reported that incorporating 10% bean flour into wheat flour for bread production enhanced the dough's nutritional, technological, and rheological properties.

**Table 2.** Functional properties of common bean-based food products

| Bioactive compound | Product | Techno-functional property | Result | References |
|---|---|---|---|---|
| Proteins | Gluten-free cookies | Water absorption capacity and viscosity. | Incorporating 40% raw pinto bean flour to produce gluten-free cookies showed excellent sensory acceptance regarding water absorption and viscosity compared to flours made from germinated and cooked seeds. | Simons and Hall (2018) |
| Proteins | Cookies | Foaming and emulsifying capacity. | Formulated cookies based on bean, chickpea, and wheat sprouts have a nutritional value of proteins (12.32%), lipids (22.57%) and fiber (5.64%), impacting functional properties such as water (1.46 g/g) and oil (1.38 g/g) absorption capacity, foaming (21.43%), and emulsifying (24.45 mL oil/g) compared to the raw material. | Sibian and Riar (2020) |
| Proteins | Baking dough | Foaming capacity | The hydration capacity of the proteins favored the formation of foam, mainly due to the presence of vicilins. | Shevkani et al. (2022) |
| Proteins and carbohydrates | Flours | Foaming agent, emulsifier, and water and oil absorption capacity. | They observed that the incorporation of 30% red bean flour in potato flour presented water absorption capacity (2.50 mL/g), oil absorption capacity (0.75 mL/g), emulsifying (7.31 mg/g), foaming (5%), and foam stability (3.70%). The mixture could be used as a basis for complementary foods. | Bossi et al. (2023) |
| Proteins | Gel-based foods | Water retention capacity. | Bean protein isolates obtained at pH 3 and 7 demonstrated that the major protein fraction is globulin (7S), also known as vicilin. This suggests that electrostatic, hydrophobic, and hydrogen bonding interactions are the primary intermolecular forces involved in gel formation. The findings indicate that these protein isolates hold potential for application in meat and dairy products due to their functional properties. | Ge et al. (2023) |

Due to their composition and techno-functional properties, beans have been considered a viable alternative to meat and pasta (Stefanucci et al., 2022). They are also increasingly being incorporated into the development of new products, such as snacks, gluten-free breakfast cereals, pasta, protein gels, soups, cookies, biscuits, spreads, and breakfast foods (Pedrosa et al., 2020; Shevkani et al., 2022; De Paiva Gouvêa et al., 2023). The techno-functional properties of beans include swelling and hydration capacity, which are influenced by factors such as particle size and the content of proteins and polysaccharides, particularly starch and dietary fiber (Table 2).

## 4.1. Microencapsulation

Resistant starch is a fraction of starch that resists digestion and reaches the colon to be fermented by the microbiota, producing short-chain fatty acids such as acetate, propionate, and butyrate that contribute to the prevention of colon cancer (Escobedo et al., 2020). Resistant starch improves food qualities such as increased crunchiness, reduced oil in fried foods, and even microencapsulation of probiotics as a shell material (Homayouni et al., 2014).

Bioactive compounds of interest in food include ingredients that contribute to flavor, color, and provide health benefits, such as antioxidants, polyunsaturated fatty acids, and polyphenols. Flavors, whether liquid or solid, are a complex mixture of volatile substances and labile components, susceptible to oxidation, chemical interactions, or volatilization and, consequently, affect sensory perception (Nicolás-García et al., 2022; Stefanucci et al., 2022; Nartea et al., 2023). To minimize this damage, microencapsulation is used to trap liquid flavoring substances: essential oils, oleoresins, flavor mixtures, and acidulants. In this regard, research has been carried out to evaluate the composition of the wall material and the operating conditions for the retention and control of the release of encapsulated flavors (Koop et al., 2022; Fabela-Morón et al., 2022; Pop et al., 2023).

Incorporating proteins as food encapsulating agents increases porosity, surface flexibility, mechanical resistance, and water evaporation (Wen et al., 2022; Jayaprakash et al., 2023). In the food industry, antioxidant compounds have become of great interest due to their potential benefits as important nutraceutical ingredients, as they have a protective effect on the oxidative process (Chávez-Mendoza et al., 2017; Loizzo and Tundis, 2019; Pop et al., 2023). The antioxidant activity of the common bean is increased after digestion because of an increased release of phenolic compounds due to the acidic environment prevailing in the stomach and enzyme-mediated hydrolysis in the duodenum (Akillioglu et al., 2010; Wang et al., 2019). The efficacy of these compounds depends on the active compound's stability, bioactivity, and bioavailability. However, environmental factors such as light, heat, and oxygen must be considered for their application in food. Encapsulation is an effective process for their preservation (Koop et al., 2022; Fabela-Morón et al., 2022) largely because of their bioactive nutrient content and importance as dietary antioxidants. Encapsulation of polyunsaturated fatty acids, such as omega-3 and omega-6 fatty acids, is commonly used in food fortification to preserve characteristics such as odor, flavor, and oxidative stability (Quintanilla-Carvajal et al., 2010; Carneiro et al., 2013; Cao et al., 2021; Khwaldia et al., 2022).

### 4.2. Cooking

Currently, all commercial varieties of bean grains are subjected to a cooking process before consumption, which increases starch digestibility and modifies nutritional components such as proteins. Textural changes during the thermal processing of starchy foods involve hydration, solubilization, and swelling of starch granules, dissolution of the cell wall, and lamellar polymers found in the cotyledon (Perea-Flores et al., 2011; Chigwedere et al., 2018). Cooking of common bean seeds results in new sensory properties such as flavor, aroma, and texture of the seed (Teixeira-Guedes et al., 2019), improving bioaccessibility during digestion (Minatel et al., 2017; Cárdenas-Castro et al., 2020).

## Conclusion

The common bean (*Phaseolus vulgaris* L.) has been an emblematic food of Mexican culture since pre-Hispanic times for its nutritional attributes. It is an excellent protein, fiber, calcium, and iron source. It has different biological activities such as antioxidant, anticarcinogenic, and antidiabetic. Due to its composition and techno-functional properties (solubility, water, and oil absorption), beans' bioactive compounds have been considered an alternative to replace meat and pasta in the food industry. This chapter presents the functional properties of bioactive compounds in common beans and their application in the food industry. As a food crop for direct consumption, beans are associated with minimal processing before reaching the consumer, playing an important role in food safety and the fight against malnutrition.

## References

Aguiló-Aguayo, I., Álvarez, C., Saperas, M., Rivera, A., Abadias, M., & Lafarga, T. (2021). Proteins isolated from Ganxet common bean (*Phaseolus vulgaris* L.) landrace: techno-functional and antioxidant properties. *International Journal of Food Science and Technology*, *56* (11), 5452-5460. doi:10.1111/ijfs.15201.

Akillioglu, H. G., & Karakaya, S. (2010). Changes in total phenols, total flavonoids, & antioxidant activities of common beans and pinto beans after soaking, cooking, and in vitro digestion process. *Food Science and Biotechnology*, *19*, 633-639. doi: 10.1007/s10068-010-0089-8.

Alcázar-Valle, M., García-Morales, S., Mojica, L., Morales-Hernández, N., Sánchez-Osorio, E., Flores-López, L., Enríquez-Vara, J. N., & Lugo-Cervantes, E. (2021). Nutritional, antinutritional compounds and nutraceutical significance of native bean species (*Phaseolus* spp.) of Mexican cultivars. *Agriculture*, *11* (11), 1031. doi:10.3390/agriculture11111031.

Alcázar-Valle, M., Lugo-Cervantes, E., Mojica, L., Morales-Hernández, N., Reyes-Ramírez, H., Enríquez-Vara, J. N., & García-Morales, S. (2020). Bioactive compounds, antioxidant activity, & antinutritional content of legumes: a comparison between four *Phaseolus* species. *Molecules*, *25* (15), 3528. doi:10.3390/molecules25153528.

Alfaro-Diaz, A., Urías-Silvas, J. E., Loarca-Piña, G., Gaytan-Martínez, M., Prado-Ramirez, R., & Mojica, L. (2021). Techno-functional properties of thermally treated black bean protein concentrate generated through an ultrafiltration process. *LWT*, *136*, 110296. doi:10.1016/j.lwt.2020.110296.

Al-Ruwaih, N., Ahmed, J., Mulla, M. F., & Arfat, Y. A. (2019). High-pressure assisted enzymatic proteolysis of kidney beans protein isolates & characterization of hydrolysates by functional, structural, rheological and antioxidant properties. *LWT*, *100*, 231-236. doi:10.1016/j.lwt.2018.10.074.

Ashraf, J., Liu, L., Awais, M., Xiao, T., Wang, L., Zhou, X., Tong, L.T, & Zhou, S. (2020). Effect of thermosonication pre-treatment on mung bean (*Vigna radiata*) and white kidney bean (*Phaseolus vulgaris*) proteins: Enzymatic hydrolysis, cholesterol lowering activity and structural characterization. *Ultrasonics Sonochemistry*, *66*, 105121. doi:10.1016/j.ultsonch.2020.105121.

Bogweh, N. E., & Ageyo, O. C. (2021). Do common beans (*Phaseolus vulgaris* L.) promote good health in humans? a systematic review and meta-analysis of clinical and randomized controlled trials. *Nutrients* 2021, 13(11), 3701. doi:10.3390/nu13113701.

Bosmali, I., Giannenas, I., Christophoridou, S., Ganos, C. G., Papadopoulos, A., Papathanasiou, F., Kolonas, A., & Gortzi, O. (2023). Microclimate and Genotype Impact on Nutritional and Antinutritional Quality of Locally Adapted Landraces of

Common Bean (*Phaseolus vulgaris* L.). *Foods*, 12(6), 1119. doi:10.3390/foods 12061119.

Bossi, D. S. D., Saah, M. B. D., Njapndounké, B., & Ngoufack, F. Z. (2023). Nutritional evaluation, oxidative indexes, and functional properties of Irish potatoes, eggs, and red kidney beans based complementary food. *The North African Journal of Food and Nutrition Research*, *7* (15), 20-30. doi:10.51745/najfnr.7.15.20-30.

Cao, H., Saroglu, O., Karadag, A., Diaconeasa, Z., Zoccatelli, G., Conte-Junior, C. A., & Xiao, J. (2021). Available technologies on improving the stability of polyphenols in food processing. *Food Frontiers*, *2*(2), 109-139. doi:10.1002/fft2.65.

Carboni, A. D., Salinas, M. V., & Puppo, M. C. (2023). Production of legume-wheat dough of optimum quality for breadmaking: essential analyses required. *Current Opinion in Food Science*, *49*, 100970. doi:10.1016/j.cofs.2022.100970.

Carneiro, H. C. F., Tonon, R. V., Grosso, C. R. F., & Hubinger, M. D. (2013). Encapsulation efficiency and oxidative stability of flaxseed oil microencapsulated by spray drying using different combinations of wall materials. *Journal of Food Engineering*, 115(4), 443–451. doi:10.1016/j.jfoodeng.2012.03.033.

Chávez-Mendoza, C., &Sánchez, E. (2017). Bioactive compounds from Mexican varieties of the common bean (*Phaseolus vulgaris*): Implications for health. *Molecules*, *22*(8), 1360. doi:10.3390/molecules22081360.

Chigwedere, C. M., Olaoye, T. F., Kyomugasho, C., Jamsazzadeh Kermani, Z., Pallares Pallares, A., Van Loey, A. M., Grauwet, T., & Hendrickx, M. E. (2018). Mechanistic insight into softening of Canadian wonder common beans (*Phaseolus vulgaris*) during cooking. *Food Research International*, *106*, 522–531. doi:10.1016/j.foodres.2018. 01.016.

Cominelli, E., Sparvoli, F., Lisciani, S., Forti, C., Camilli, E., Ferrari, M., Le Donne, C., Marconi, S., Juan Vorster, B., Botha, A. M., Marais, D., Losa, A., Sala, T., Reboul, E., Alvarado-Ramos, K., Waswa, B., Ekesa, B., Aragão, F., & Kunert, K. (2022). Antinutritional factors, nutritional improvement, and future food use of common beans: A perspective. *Frontiers in Plant Science*, 13(August), 1–8. doi:10.3389/fpls. 2022.992169.

Corzo-Ríos, L. J., Sánchez-Chino, X. M., Cardador-Martínez, A., Martínez-Herrera, J., & Jiménez-Martínez, C. (2020). Effect of cooking on nutritional and non-nutritional compounds in two species of Phaseolus (*P. vulgaris* and *P. coccineus*) cultivated in Mexico. *International Journal of Gastronomy and Food Science*, 20(October 2019), 100206. doi:10.1016/j.ijgfs.2020.100206.

David I, Orboi MD, Simandi MD, Chirilă CA, Megyesi CI, Rădulescu L, et al. Fatty acid profile of Romanian's common bean (*Phaseolus vulgaris* L.) lipid fractions and their complexation ability by β-cyclodextrin. *PLoS One*. (2019) 14: e0225474. doi:10.1371/journal.pone.0225474.

Demiate, I. M., Figueroa, A. M., Zortéa Guidolin, M. E. B., Rodrigues dos Santos, T. P., Yangcheng, H., Chang, F., & Jane, J.-l. (2016). Physicochemical characterization of starches from dry beans cultivated in Brazil. *Food Hydrocolloids*, *61*, 812-820. doi:10.1016/j.foodhyd.2016.07.014.

De Paiva Gouvêa, L., Caldeira, R., de Lima Azevedo, T., Galdeano, M. C., Felberg, I., Lima, J. R., & Mellinger, C. G. (2023). Physical and techno-functional properties of a common bean protein concentrate compared to commercial legume ingredients for the plant-based market. *Food Hydrocolloids*, *137*, 108351. doi:10.1016/j.foodhyd. 2022.108351.

Duenas, M., Sarmento, T., Aguilera, Y., Benitez, V., Molla, E., Esteban, R. M., & Martín-Cabrejas, M. A. (2016). Impact of cooking and germination on phenolic composition and dietary fibre fractions in dark beans (*Phaseolus vulgaris* L.) and lentils (*Lens culinaris* L.). *LWT-Food Science and Technology*, *66*, 72-78. doi:10.1016/j.lwt.2015.10.025.

Fabela-Morón, M. F., Pérez-Ruíz, R. V., Ruíz-Hernández, R., Arce-Vázquez, M. B., Aguilar-Toalá, J. E., Jiménez-Guzmán, J., & García-Garibay, J. M. (2022). Encapsulation of bioactive compounds of food interest: applications, current advances, challenges, and opportunities. *Agro Productividad*. doi:10.32854/ agrop.v15i10.2407.

Finetti, F., Biagi, M., Ercoli, J., Macrì, G., Miraldi, E., & Trabalzini, L. (2020). *Phaseolus vulgaris* L. var. Venanzio grown in Tuscany: Chemical composition and in vitro investigation of potential effects on colorectal cancer. *Antioxidants*, *9* (12), 1181. doi:10.3390/antiox9121181.

Ganesan, K., & Xu, B. (2017). Polyphenol-rich dry common beans (*Phaseolus vulgaris* L.) and their health benefits. *International journal of molecular sciences*, *18*(11), 2331. doi:10.3390/ijms18112331.

Gani, A., Bashir, M., Wani, S. M., & Masoodi, F. A. (2012). Modification of bean starch by γ-irradiation: Effect on functional and morphological properties. *LWT*, *49*(1), 162-169. doi:10.1016/j.lwt.2012.04.028.

Ge, J., Sun, C., Chang, Y., Li, S., Zhang, Y., & Fang, Y. (2023). Understanding the differences in heat-induced gel properties of twelve legume proteins: A comparative study. *Food Research International*, *163*, 112134. doi:10.1016/j.foodres. 2022.112134.

Goñi, I., Díaz-Rubio, M. E., Pérez-Jiménez, J., & Saura-Calixto, F. (2009). Towards an updated methodology for measurement of dietary fiber, including associated polyphenols, in food and beverages. *Food Research International*, *42*(7), 840-846. doi:10.1016/j.foodres.2009.03.010.

Grela, E. R., & Günter, K. D. (1995). Fatty acid composition and tocopherol content of some legume seeds. *Animal feed science and technology*, *52*(3-4), 325-331. doi:10.1016/0377-8401(94)00733-P.

Hernández-Guerrero, C. J., Villa-Ruano, N., Zepeda-Vallejo, L. G., Hernández-Fuentes, A. D., Ramirez-Estrada, K., Zamudio-Lucero, S., Hidalgo-Martínez, D., & Becerra-Martínez, E. (2021). Bean cultivars (*Phaseolus vulgaris* L.) under the spotlight of NMR metabolomics. *Food Research International*, 150, 110805. doi:10.1016/j.foodres.2021.110805.

Homayouni, A., Amini, A., Keshtiban, A. K., Mortazavian, A. M., Esazadeh, K., & Pourmoradian, S. (2014). Resistant starch in food industry: A changing outlook for consumer and producer. *Starch-Stärke*, *66*(1-2), 102-114. doi:10.1002/star. 201300110.

Jayaprakash, P., Maudhuit, A., Gaiani, C., & Desobry, S. (2022). Encapsulation of bioactive compounds using competitive emerging techniques: Electrospraying, nano spray drying, and electrostatic spray drying. *Journal of Food Engineering*, 111260. doi:10.1016/j.jfoodeng.2022.111260.

Khwaldia, K., Attour, N., Matthes, J., Beck, L., & Schmid, M. (2022). Olive byproducts and their bioactive compounds as a valuable source for food packaging applications. *Comprehensive Reviews in Food Science and Food Safety*, 21(2), 1218–1253. doi:10.1111/1541-4337.12882.

Koop, B. L., Nascimento da Silva, M., Diniz da Silva, F., Thayres dos Santos Lima, K., Santos Soares, L., José de Andrade, C., Ayala Valencia, G., & Rodrigues Monteiro, A. (2022). Flavonoids, anthocyanins, betalains, curcumin, and carotenoids: Sources, classification and enhanced stabilization by encapsulation and adsorption. *Food Research International*, 153(January), 110929. doi:10.1016/j.foodres.2021.110929.

Kulaz, H., Najafi, S., Tuncturk, R., Tuncturk, M., Albalawi, M. A., Alalawy, A. I., Oyouni, A. A. A., Alasmari, A., Poczai, P., & Sayyed, R. Z. (2023). Analysis of Nuclear DNA Content and Karyotype of *Phaseolus vulgaris* L. *Genes*, 14(1). doi:10.3390/genes14010047.

Lima-Becerra, I., María, B. A., Dorantes-Campuzano, F., Mojica, L., Loarca-Piña, G., Morales-Sánchez, E., Ramírez-Jiménez, A. K., & Gaytán-Martínez, M. (2021). Ohmic heating as an emerging technology for the improvement of the techno-functional properties of common bean flour. *In Biology and Life Sciences Forum*, *6* (1), 95. doi:10.3390/Foods2021-11000.

Loizzo, M. R., & Tundis, R. (2019). Plant antioxidant for application in food and nutraceutical industries. *Antioxidants*, *8*(10), 453. doi:10.3390/antiox8100453.

Alcázar-Valle, M., Lugo-Cervantes, E., Mojica, L., Morales-Hernández, N., Reyes-Ramírez, H., Enríquez-Vara, J. N., & García-Morales, S. (2020). Bioactive compounds, antioxidant activity, and antinutritional content of legumes: a comparison between four phaseolus species. *Molecules*, *25*(15), 3528. doi:10.3390/molecules 25153528.

Martínez-Ávila, M., Rodríguez-Rodríguez, J., Uribe, J. G., & Guajardo-Flores, D. (2022). Selective supercritical fluid extraction of non-polar phytochemicals from black beans (*Phaseolus vulgaris* L.) by-products. *The Journal of Supercritical Fluids*, *189*, 105730. doi:10.1016/j.supflu.2022.105730.

Mikić, A. (2018). *Phaseolus* L. In Lexicon of Pulse Crops (pp. 187–211). doi:10.1201/b22282-13.

Nartea, A., Kuhalskaya, A., Fanesi, B., Orhotohwo, O. L., Susek, K., Rocchetti, L., Di Vittori, V., Bitocchi, E., Pacetti, D., & Papa, R. (2023). Legume by-products as ingredients for food applications: Preparation, nutrition, bioactivity, and techno-functional properties. *Comprehensive Reviews in Food Science and Food Safety*. doi:10.1111/1541-4337.13137.

Nicolás-García, M., Perucini-Avendaño, M., Arrieta-Báez, D., de Jesús Perea-Flores, M., Jiménez-Martínez, C., Gómez-Patiño, M. B., & Dávila-Ortiz, G. (2022). Phenolic compounds profile by UPLC-ESI-MS in black beans and its distribution in the seed coat during storage. *Food Chemistry*, 133638. doi:10.1016/j.foodchem.2022.133638.

Nicolás-García, M., Perucini-Avendaño, M., Jiménez-Martínez, C., Perea-Flores, M. D. J., Gómez-Patiño, M. B., Arrieta-Báez, & D., Dávila-Ortiz, G. (2021). Bean phenolic compound changes during processing: Chemical interactions and identification. *Journal of Food Science*, *86* (3), 643-655. doi:10.1111/1750-3841.15632.

Pedrosa, M. M., Varela, A., Domínguez-Timón, F., Tovar, C. A., Moreno, H. M., Borderías, A. J., & Díaz, M. T. (2020). Comparison of bioactive compounds content and techno-functional properties of pea and bean flours and their protein isolates. *Plant Foods for Human Nutrition*, *75*, 642-650. doi:10.1007/s11130-020-00866-4.

Perea-Flores, M. J., Chanona-Perez, J. J., Garibay-Febles, V., Calderon-Dominguez, G., Terrés-Rojas, E., Mendoza-Perez, J. A., & Herrera-Bucio, R. (2011). Microscopy techniques and image analysis for evaluation of some chemical and physical properties and morphological features for seeds of the castor oil plant (*Ricinus communis*). *Industrial Crops and Products*, *34*(1), 1057-1065. doi:10.1016/j.indcrop.2011.03.015.

Perera, D., Devkota, L., Garnier, G., Panozzo, J., & Dhital, S. (2023). Hard-to-cook phenomenon in common legumes: chemistry, mechanisms and utilisation. *Food Chemistry*, 135743. doi:10.1016/j.foodchem.2023.135743.

Pop, O. L., Suharoschi, R., Socaci, S. A., Berger Ceresino, E., Weber, A., Gruber-Traub, C., & Johansson, E. (2023). Polyphenols—Ensured Accessibility from Food to the Human Metabolism by Chemical and Biotechnological Treatments. *Antioxidants*, *12*(4), 865. doi:10.3390/antiox12040865.

Quintanilla-Carvajal, M. X., Camacho-Díaz, B. H., Meraz-Torres, L. S., Chanona-Pérez, J. J., Alamilla & Beltrán, L., Jimenéz-Aparicio, A., and Gutiérrez-López, G. F. (2010). Nanoencapsulation: A New Trend in Food Engineering Processing. *Food Engineering Reviews*, 2(1), 39–50. doi:10.1007/s12393- 009-9012-6.

Ramírez-Jiménez, A. K., Gaytán-Martínez, M., Morales-Sánchez, E., & Loarca-Piña, G. (2018). Functional properties and sensory value of snack bars added with common bean flour as a source of bioactive compounds. *LWT*, *89*, 674-680. doi:10.1016/j.lwt.2017.11.043.

Recek, N., Holc, M., Vesel, A., Zaplotnik, R., Gselman, P., Mozetič, M., & Primc, G. (2021). Germination of Phaseolus vulgaris L. seeds after a short treatment with a powerful RF plasma. *International journal of molecular sciences*, *22*(13), 6672. doi:10.3390/ijms22136672.

Rodríguez, L., Mendez, D., Montecino, H., Carrasco, B., Arevalo, B., Palomo, I., & Fuentes, E. (2022). Role of *Phaseolus vulgaris* L. in the prevention of cardiovascular diseases-cardioprotective potential of bioactive compounds. *Plants*, 2022, *11*(2), 186. doi:10.3390/plants11020186.

Sánchez-Chino, X. M., Corzo-Rios, L. J., Cid-Gallegos, M. S., & Jiménez-Martínez, C (2022). Compuestos antioxidantes provenientes de leguminosas como agentes quimiopreventivos en cáncer de colon. *Tequio*, *5* (15), 111–124.

Santos, T. D., & Furlong, E. B. (2022). Biological contamination of the common bean (*Phaseolus vulgaris* L.) and its impact on food safety. *Critical Reviews in Food Science and Nutrition*, 62(18), 4998–5004. doi:10.1080/10408398.2021.1881038.

Shamseldin, A., & Velázquez, E. (2020). The promiscuity of *Phaseolus vulgaris* L. (common bean) for nodulation with rhizobia: a review. *World Journal of Microbiology and Biotechnology*, 36(5), 1–12. doi:10.1007/s11274-020-02839-w.

Shevkani, K., Kaur, R., Singh, N., & Hlanze, D. P. (2022). Colour, composition, digestibility, functionality and pasting properties of diverse kidney beans (*Phaseolus vulgaris*) flours. *Current Research in Food Science*, *5*, 619-628. doi:10.1016/j.crfs.2022.03.006.

Sibian, M. S., & Riar, C. S. (2020). Formulation and characterization of cookies prepared from the composite flour of germinated kidney bean, chickpea, and wheat. *Legume Science*, *2* (3), e42. doi:10.1002/leg3.42.

Simons, C. W., & Hall III, C. (2018). Consumer acceptability of gluten-free cookies containing raw cooked and germinated pinto bean flour. *Food Science and Nutrition*, *6* (1), 77-84. doi:10.1002/fsn3.531.

Stefanucci, A., Scioli, G., Marinaccio, L., Zengin, G., Locatelli, M., Tartaglia, A., Della Valle, A., Cichelli, A., Novellino, E., Pieretti, S., & Mollica, A. (2022). A comparative study on phytochemical fingerprint of two diverse *Phaseolus vulgaris* var. Tondino del Tavo and Cannellino Bio extracts. *Antioxidants*, *11* (8), 1474. doi:10.3390/antiox11081474.

Teixeira-Guedes, C. I., Oppolzer, D., Barros, A. I., & Pereira-Wilson, C. (2019). Impact of cooking method on phenolic composition and antioxidant potential of four varieties of *Phaseolus vulgaris* L. and *Glycine max* L. *LWT*, *103*, 238–246. doi:10.1016/ j.lwt.2019.01.010.

Venegas, G. R., Ita, D. L. De, Díaz, P. M., Méndez, M. I., Gasca, T. G., & Solórzano, M. A. (2021). Supplementation with *Phaseolus vulgaris* Leaves Improves Metabolic Alterations Induced by High - Fat / Fructose Diet in Rats Under Time - Restricted Feeding. *Plant Foods for Human Nutrition*, 297–303. doi:10.1007/s11130-021-00904-9.

Vogelsang-O'Dwyer, M., Sahin, A. W., Arendt, E. K., & Zannini, E. (2022). Enzymatic hydrolysis of pulse proteins as a tool to improve techno-functional properties. *Foods*, *11* (9), 1307. doi:10.3390/foods11091307.

Wainaina, I., Wafula, E., Sila, D., Kyomugasho, C., Grauwet, T., Van Loey, A., & Hendrickx, M. (2021). Thermal treatment of common beans (*Phaseolus vulgaris* L.): Factors determining cooking time and its consequences for sensory and nutritional quality. *Comprehensive Reviews in Food Science and Food Safety*, *20* (4), 3690-3718. doi:10.1111/1541-4337.12770.

Wang, C., & Xie, Y. (2019). Interaction of protein isolate with anthocyanin extracted from black soybean and its effect on the anthocyanin stability. *Journal of Food Science*, *84*(11), 3140-3146. doi:10.1111/1750-3841.14816.

Wen, C., Zhang, J., Zhang, H., & Duan, Y. (2022). New perspective on natural plant protein-based nanocarriers for bioactive ingredients delivery. *Foods*, *11*(12), 1701. doi:10.3390/foods11121701.

Xu, B., & Chang, S. K. (2009). Total phenolic, phenolic acid, anthocyanin, flavan-3-ol, and flavonol profiles and antioxidant properties of pinto and black beans (*Phaseolus vulgaris* L.) as affected by thermal processing. *Journal of Agricultural and Food Chemistry*, *57*(11), 4754-4764. doi:10.1021/jf900695s.

Yang, Q. Q., Gan, R. Y., Ge, Y. Y., Zhang, D., & Corke, H. (2018). Polyphenols in common beans (*Phaseolus vulgaris* L.): Chemistry, analysis, and factors affecting composition. *Comprehensive Reviews in Food Science and Food Safety*, *17*(6), 1518-1539. doi:10.1111/1541-4337.12391.

## Chapter 3

# Chickpea *(Cicer arietinum)*: Bioactive Compounds

**Z. E. Olivo-Vidal[1]**
**A. Linares-Castañeda[2]**
**L. J. Corzo-Ríos[3]**
**M. S. Cid-Gallegos[3]**
**C. Jiménez-Martínez[2]**
**and X. M. Sánchez-Chino[1,*]**

[1]Departamento de Salud, El Colegio de la Frontera Sur-Villahermosa, Villahermosa, Tabasco, México
[2]Escuela Nacional de Ciencias Biológicas, Instituto Politécnico Nacional, Ciudad de México, México
[3]Unidad Profesional Interdisciplinaria de Biotecnología, Instituto Politécnico Nacional, Ciudad de México, México

### Abstract

The chickpea (*Cicer arietinum)* belongs to the family of legumes and seeds that are a source of protein worldwide in nutritional terms for people who have decided not to eat foods of animal origin and those without the financial means to get them. In addition, it contains other nutrients such as carbohydrates, lipids, fiber, and minerals. These seeds present non-nutritional compounds, also known as bioactive compounds, such as phenolic compounds (flavonoids, anthocyanins, isoflavones), phytates, alpha galactosides, trypsin inhibitors, and peptides, which have been reported to be beneficial for health because they have

* Corresponding Author's Email: xsanchez@ecosur.mx

In: Legumes: Beyond Their Nutritional Value
Editors: C. Jiménez-Martínez, M. F. Juárez-Chairez and M. S. Cid-Gallegos
ISBN: 979-8-89530-440-2

antihypertensive, antioxidant, anticancer, and anticholesterolemic properties. Thus, their consumption alone or through the consumption of the whole seed could help prevent diseases of public health relevance.

**Keywords:** chickpea, nutrients, proteins, carbohydrates, lipids, bioactive compounds

## 1. Introduction

The legume family (*Fabaceae* or *Leguminosae*) is the third-largest family of angiosperms, with approximately 750 genera and 20,000 species (Mathesius, 2022; Tas & Shah, 2021). The presence of proteins and starches in adequate proportions and their content of fiber, vitamins, minerals, and bioactive compounds have been studied due to the multiple advantages that consuming them provides (Muzquiz et al., 2012). Globally, some of the most widely consumed grain legumes include beans (*Phaseolus* spp.), broad beans (*Vicia* spp.), peas (*Pisum* spp.), chickpeas (*Cicer arietinum*), cowpeas (*Vigna unguiculata*), pigeon peas (*Cajanus cajan*), lentils (*Lens culinaris*), lupines (*Lupinus* spp.), among others (Tas & Shah, 2021).

Chickpea is one of the world's most cultivated and consumed legumes, mainly in tropical and subtropical areas. They are cultivated mainly in the Mediterranean, the Near East, Central Asia, and America (Saleh et al., 2006). In these areas, chickpeas are the basis for the preparation of various foods, especially in the Mediterranean and Middle Eastern areas, such as hummus, as well as in India, where chickpea-based foods are prevalent, such as boondi and dhokla (Figure 1), snacks, salads, and different stews based on chickpeas and meat from different origins (Kaur et al., 2021); hence, this seed represents one of the most important legumes in nutrition worldwide, representing 17% of the global production of legumes (Acevedo-Martínez & González de Mejía, 2021).

Chickpea seeds (Figure 2) are rich in carbohydrates (70%) and proteins (19%); they also contain small amounts of lipids (6%), fiber (1.7%), and minerals (3.4%). In addition, they contain compounds that are part of the secondary metabolites generated by plants, many of which are a form of defense against environmental aggressions and are known as non-nutritional. These compounds do not contribute to caloric intake, but their consumption provides significant health benefits, which is why they are referred to as bioactive compounds. Among these bioactive compounds are phenolic compounds, isoflavones, saponins, phytates, protease inhibitors, lectins, and α-galactosides (Primi et al., 2019; Sánchez-Chino et al., 2017).

**Table 1.** Nutritional composition of chickpea varieties

| | Desi | Kabuli | Reference |
|---|---|---|---|
| Macronutrients (g/100g) | | | |
| Moisture | 7.89 | 7.66 | Xiao et al., 2023 |
| | 8.6 | - | Sofi et al., 2023 |
| Protein | 22.50 * | 19.79 | Xiao et al., 2023 |
| | 25.92 * | 21.38 * | Costantini et al., 2021 |
| Lipids | 6.65 | 8.39 * | Xiao et al., 2023 |
| | 3.41 * | 4.47 * | Costantini et al., 2021 |
| Starch | 27.15 | 36.21 | Xiao et al., 2023 |
| | 38.48 * | 39.12 * | Rachwa-Rosiak & Nebesny, 2015 |
| Total dietary fiber | 18.15 * | 7.94 * | Costantini et al., 2021 |
| Micronutrients (mg/100g) | | | |
| Ash | 2.66 | 2.66 * | Xiao et al., 2023 |
| Ca | 165.0 * | 81.7 * | Rachwa-Rosiak & Nebesny, 2015 |
| | 210.17 * | 174.71 * | Costantini et al., 2021 |
| K | 994.5 * | 1060.0 * | Rachwa-Rosiak & Nebesny, 2015 |
| | 1007.91 * | 957.08 * | Costantini et al., 2021 |
| Mg | 169.0 * | 147.0 * | Rachwa-Rosiak & Nebesny, 2015 |
| | 230.52 | 197.73 | Xiao et al., 2023 |
| Fe | 4.59 * | 5.50 * | Rachwa-rosiak & Nebesny, 2015 |
| | 9.72 | 7.36 | Xiao et al., 2023 |
| P | 451.5 * | 394.0 * | Rachwa-rosiak & Nebesny, 2015 |
| | 395.15 * | 386.94 * | Costantini et al., 2021 |
| Zn | 4.07 * | 3.40 * | Rachwa-rosiak & Nebesny, 2015 |
| | 4.47 | 3.61 | Xiao et al., 2023 |
| Mn | 3.81 * | 3.28 * | Rachwa-Rosiak & Nebesny, 2015 |
| | 3.32 * | 3.26 * | Costantini et al., 2021 |
| Vitamins (mg/100 g) | | | |
| Thiamine (B1) | 0.31 | 0.36 | Xiao et al., 2023 |
| | 0.4 | 0.29 | Wang et al., 2021 |
| Riboflavin (B2) | 0.21 | 0.26 | Wang et al., 2021 |
| Niacin (B3) | 1.72 | 1.22 | Wang et al., 2021 |
| Retinol (A) | ND | ND | Wang et al., 2021 |
| Vitamin C | 1.65 | 1.34 | Wang et al., 2021 |
| Folic acid (B9) | 0.206 | 0.299 | Wang et al., 2021 |

* dry base.

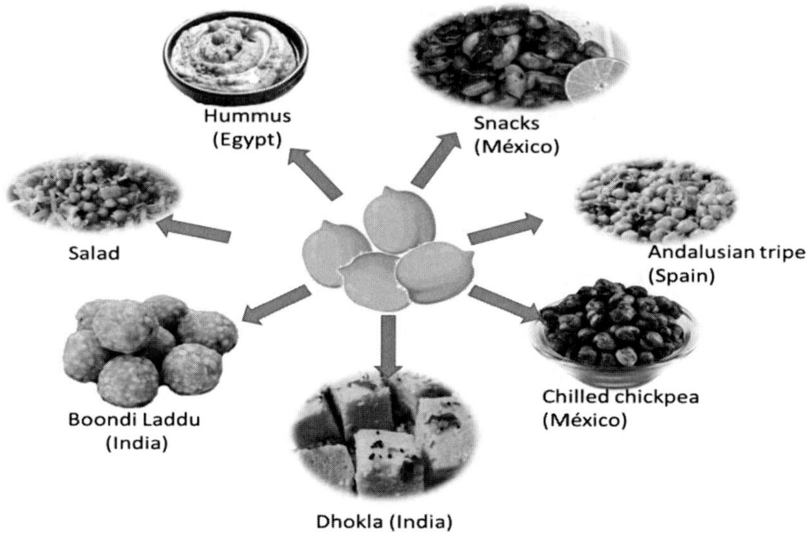


**Figure 1.** Foods made with chickpeas.

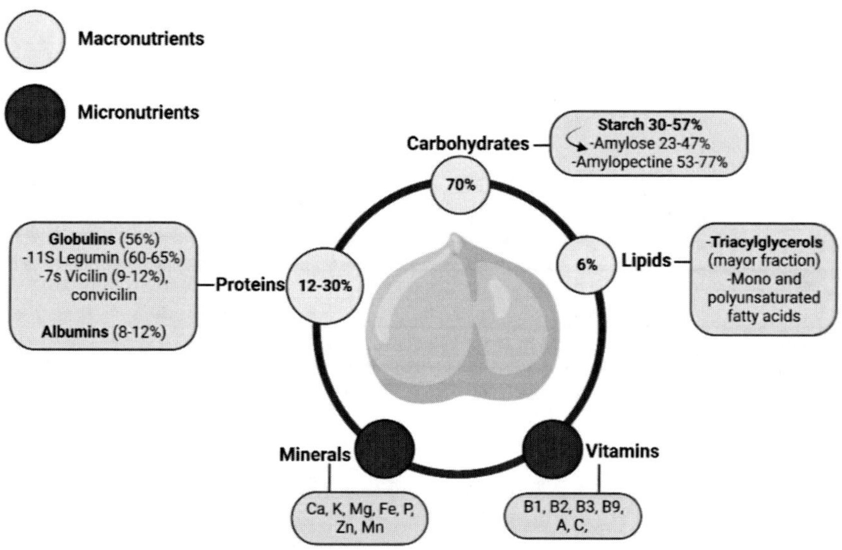


**Figure 2.** Macro and micronutrients are present in chickpeas.

The number of macronutrients, micronutrients, and bioactive compounds in terms of concentration and diversity will depend on the type of seed and variety (Table 1). In the case of chickpeas, two varieties are identified: Kabuli, which has large beige seeds, and Desi, which has more minor and rougher brown to black seeds (Wood et al., 2011). Both are cultivated under the same environmental and agronomic conditions (Primi et al., 2019; Pasqualone et al., 2021), postharvest handling (Berhe et al., 2024) and are prepared similarly for final consumption (Quintero-Hilario et al., 2019, Cardador-Martínez et al., 2020; Corzo-Ríos et al., 2020, Corzo-Ríos et al., 2022).

In general, mature chickpea seeds have three important morphological parts: the coat or cover (outer layer), the cotyledons (intermediate), and the embryonic axis (inner) (González, 1987), among which the chemical composition can vary in terms of the concentration of macronutrients (mainly highlighting proteins, carbohydrates, and lipids), micronutrients, and components considered non-nutrients (bioactive compounds).

## 2. Macronutrients

### 2.1. Proteins

The protein content in chickpea seeds ranges from 12% to 30%, though it is deficient in sulfur-rich amino acids, with albumins and globulins being the most abundant protein fractions (Wallace et al., 2016). Proteins are essential nutrients required for cell turnover and various other health-promoting functions, depending on their amino acid composition and structure. In addition to their nutritional roles, catalytic proteins play key functions in metabolism, and some can act as non-nutritional compounds by inhibiting proteolytic enzymes. One such example is protease inhibitors, which are low-molecular-weight proteins with enzymatic activity. In plants, these inhibitors serve as a defense mechanism against insects and pathogens (Hellinger & Gruber, 2019). In humans, protease inhibitors can affect health both directly and indirectly. For instance, they can enhance the absorption of protein-based drugs, such as insulin, by preventing degradation from endogenous proteases (Cid-Gallegos et al., 2022). Furthermore, protease inhibitors can directly inhibit proteases in carcinogenic cells, depriving them of oxygen and nutrients, which can slow the progression of cancer (Eatemadi et al., 2017).

On the other hand, when consumed, proteins must undergo hydrolysis for digestion. Endogenous proteases, such as pepsin and pancreatin, cleave specific peptide bonds, releasing peptides and amino acids. The health benefits of these released peptides and amino acids depend on their R groups, configuration, and size, and they can serve as potential therapeutic agents for treating various diseases. For example, in the case of chickpea seed, different peptides have been obtained that have been shown to have antihypertensive (Chávez-Ontiveros et al., 2022), antioxidant, antiproliferative (Quintero-Soto et al., 2021), anticancer (Magee et al., 2012) and hypocholesterolemic (Yust et al., 2012), and other effects. Although these bioactive peptides can be released within the normal digestion process, there are technological methods, like *in vitro* enzymatic hydrolysis, where the release can be controlled or through germination or fermentation (Ulug et al., 2021).

The activity most reported for chickpea peptides and protein hydrolysates is the antioxidant through different mechanisms: radical trapping ABTS (2,2'-azino-bis(3-ethylbenzothiazoline-6-sulfonic acid) and DPPH (2,2-Diphenyl-1-picrylhydrazyl), chelation of metals such as copper and iron, and inhibition of beta-carotene oxidation, which mainly depends on the size of the peptides and their amino acid sequence (Megías et al., 2007; Medina-Godoy et al., 2012; Yust et al., 2012; remarkably, in the copper chelating activity and β-carotene bleaching, it has been proposed that the histidine content (19.7 to 35.1%) is determinant (Megías et al., 2007). In addition, antioxidant activity is related to lipid-lowering, antihypertensive, and anticarcinogenic activities (Lorenzo et al., 2018).

From hardened chickpea seeds, Medina-Godoy et al. (2012) obtained concentrates and hydrolysates of their proteins sequentially for 2 h, using the enzymes alcalase, pancreatin, and papain, reaching a degree of hydrolysis of 8.03 and 15.59%. With this, the number of polar and hydrophilic groups was increased, modifying the globular structure of the proteins. The peptides contained in the hydrolysates presented antihypertensive activity by inhibiting the angiotensin-converting enzyme (ACE-I), showing $IC_{50}$ values between 0.101 and 37.33 μg $mL^{-1}$. The ACE-1 is relevant because it belongs to the renin-angiotensin-aldosterone system, an important regulator of cardiovascular and renal functions. Blocking this enzyme reduces blood pressure values, reducing cardiovascular risk and slowing the progression of renal failure (Khurana & Goswami, 2022).

Peptides with hypocholesterolemic capacity have also been obtained by hydrolysis with alcalase and Flavourzyme® for 60 min. The peptides released with these enzymes are rich in hydrophobic amino acids such as Ile, Asn, Tre, and Gly. These peptides were tested in an animal model of mice fed high-fat diets, observing an increase in the excretion of lipids in feces. The authors attribute this to the fact that these peptides compete with lipids to form the micelles formed for their transportation and solubilization, thus inhibiting their transport and subsequent metabolism, resulting in the excretion of lipids (Yust et al., 2012).

The potential of bioactive peptides derived from chickpeas through enzymatic hydrolysis using proteases such as pepsin, trypsin, chymotrypsin, alcalase, Flavourzyme®, and papain has been demonstrated for managing chronic non-communicable diseases, including type 2 diabetes mellitus. By inhibiting dipeptidyl peptidase (DPP-IV) inhibitors that are key regulators of postprandial insulin release by inactivating the activity of the DPP-IV enzyme and increasing the glucagon-like peptide 1 (GLP-1) involved in the control of blood glucose concentration Recent studies using *in silico* methods, have predicted DPP-IV inhibition by chickpea peptides (Real Hernández & González de Mejía, 2019). Another study, where the activity of chickpea hydrolysates with pepsin-pancreatin or bromelain was evaluated, identified that catalytic sites with high affinity for DPP-IV were generated; the sequences identified with these activities were PHPATSGGGL and YVDGSGTPLT (Chandrasekaran et al., 2020).

## 2.2. Carbohydrates

As previously mentioned, carbohydrates are the most abundant component of chickpea seeds (70%), turning them into an essential source of carbohydrates, which are, nutritionally, the main energy source. The main classes of carbohydrates present in the chickpea seed are monosaccharides like glucose (0.7%), fructose (0.25%), ribose (0.1%), and galactose (0.05%); disaccharides like sucrose (1-2%) and maltose (0.6%) and alpha-galactosides or oligosaccharides like raffinose (2.2%), stachyose 3.1%, verbascose 0.4%, and ciceritol (6.5%), the latter being the most abundant oligosaccharide in chickpea seeds, hence the genus name Cicer, and finally polysaccharides like starch, which represents 27-50% of the total carbohydrates (Wood & Grusak, 2007; Jukanti et al., 2012) and is the main source of dietary energy derived from this seed. Starch comprises two types of glucose polymers, amylose, and amylopectin, which are synthesized and form the 'granule' of starch, but each has different properties. Amylose is essentially a linear molecule corresponding to 23-47% of chickpea seeds, whereas the remaining percentage corresponds to amylopectin (53-77%), a highly branched molecule. The starch of legumes is less digestible, possibly due to the lower amylopectin content. Thus, they are generally recognized as complex carbohydrates because they are slowly absorbed and digested in the body due to their structures (Wood & Grusak, 2007).

Due to the above characteristics, these carbohydrates have been classified as low glycemic index foods. Therefore, their consumption has been recommended for the management of type 2 diabetes as they maintain stable glucose levels for a sustained time, as well as for obesity control as they increase the satiety signal; in addition, they modulate the intestinal microbiome, acting as prebiotics and improving human intestinal health (Roberfroid et al., 2009; Madurapperumage et al., 2021). It has also been suggested that a portion of the starch in legumes is resistant to digestion, often referred to as the indigestible fraction. This indigestible fraction has been associated with anticancer activity, primarily through mechanisms such as cell cycle arrest and apoptosis. The protective effects of this fraction may involve gene-nutrient interactions and the modulation of protein expression, contributing to its potential role in cancer prevention (Campos-Vega et al., 2013).

## 2.3. Lipids

Chickpea seeds contain a low proportion of lipids (6%), highlighting triacylglycerols (the majority fraction); they also contain polyunsaturated and monounsaturated fatty acids (Jukanti et al., 2012). These fatty acids are necessary for human nutrition because they are precursors for the biosynthesis of hormones and maintain cell integrity (Di Pasquale, 2009). In addition, chickpeas also contain sterols and tocopherols, known as glycerophospholipid phytosterols and glycolipid sphingolipids as membrane lipids (Zia-Ul-Haq et al., 2007).

Another element that supports the idea that consuming foods of vegetable origin, such as chickpeas, is considered beneficial for human health is the presence of unsaturated fatty acids, which can maintain cholesterol at healthy levels; hence, they also have a relevant role in preventing cardiovascular diseases. In this sense, these fatty acids can inhibit the angiotensin-converting enzyme, so their frequent consumption could have an antihypertensive effect (Ogawa et al., 2009; Kaur and Prasad, 2021).

On the other hand, it is known that when polyunsaturated fatty acids are metabolized, precursors of prostanoids and leukotrienes, such as arachidonic and eicosapentaenoic acids, are generated. Both prostanoids and leukotrienes have relevant effects on health, involving blood coagulation, bone metabolism, development and growth of the nervous system, cytoprotection of the gastrointestinal tract, secretion of mucus, fluids, and duodenal bicarbonate, tissue repair, renal function, vascular tone, and the immune and inflammatory response. These compounds are particularly important for women of reproductive age, where they are involved in ovulation and the initiation of pregnancy (Singh, 2005; Honda et al., 2023).

## 3. Bioactive Compounds

Chickpea seeds, in addition to containing nutritional compounds such as those already mentioned above, also contain non-nutritional compounds from synthesized plants in their secondary metabolism and serve as a defense to attract pollinators, among others. Their synthesis and concentration also relate to the variety (Table 2). These compounds do not provide calories. However, their consumption sometimes benefits health, so they have been called bioactive compounds. In addition, the seeds serve as defense and storage for when the metabolic machinery is activated, and new seedlings develop. Among these compounds are phenols that include flavonoids, anthocyanins, isoflavones, saponins, phytates, and alpha galactosides, among others (Figure 3).

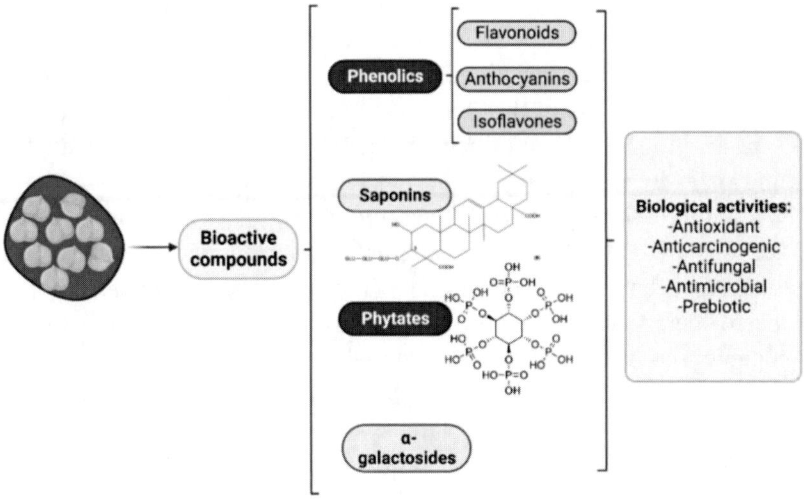


**Figure 3**. Bioactive compounds in chickpea. Created in BioRender.

**Table 2.** Bioactive compounds in chickpea varieties

| Non-nutritional | Desi | Kabuli | Reference |
|---|---|---|---|
| Total phenols (mg GAE/100 g Ds) | | 359.9 | Arevalo et al., 2020 |
| | 12.78 | 52.8 | Magalhães et al., 2017 |
| Flavonoids (mg/kg Ds) | 20.8 | 0.0 | Magalhães et al., 2017 |
| alpha galactosides (mg/100 g Ds) | | 38.13 | Arevalo et al., 2020 |
| Phytates (mg/100 g) | 1065 | | Sofi et al., 2023 |
| Totals phytates (mg phytic acid/g Ds) | 11.95 | 10.35 | Costantini et al., 2021 |
| Tannins (mg TAE/100 g) | 233 | | Sofi et al., 2023 |
| Saponins (μg/g) | 1417 | 453 | Milán-Noris et al., 2023 |

Ds= Dry base, GAE= gallic acid equivalents, TAE= tannic acid equivalents.

### 3.1. Phenols, Flavonoids, and Anthocyanins

It has been reported that chickpea seeds contain approximately 0.48 to 1.18 mg/g of phenolic compounds. These compounds are mainly present in the pericarp of the source, with ferulic acid equivalents ranging from 0.18 to 1.18 mg/g. Through the use of ultra-performance liquid chromatography coupled with mass-mass spectrometry (UPLC-MS/MS), several flavonoids and isoflavonoids have been identified, such as salicylic acid, a sinapic isomer, and protocatechuic acid glucoside. These compounds have antioxidant and anti-inflammatory properties, and they exert different mechanisms. For instance, they inhibit DNA's Nuclear Factor κB (NF-κB) binding activity, and myeloid differentiation factor 88, a critical component of the toll-like receptor pathway. This pathway is associated with the immune response and an activator of the proto-oncogene. They also suppress the release of Interleukin (IL)-6, IL-1β, IL-18, and tumor necrosis factor-alpha (TNF-α) via the NF-κB signaling pathway. Additionally, they increase messenger Ribonucleic acid (mRNA) and protein expression of Nrf2-regulated genes and inhibit the toll-like receptor 4 (TLR4). Hydroxybenzoic acid, which has been reported to have antimutagenic properties in the colon by protecting the intestinal barrier, has also been identified. Moreover, kaempferol has been linked to anti-carcinogenic activities, such as downregulation of markers related to the epithelial-mesenchymal transition (EMT) and the phosphoinositide 3-kinase/protein kinase B signaling pathways (Llorach et al., 2019; Iram et al., 2019; Alam et al., 2020; Pasqualone et al., 2021; Han et al., 2022).

Llorach et al. (2029) made a groundbreaking discovery by identifying the presence of taxifolin. This compound has been linked to the production of butyric and isobutyric acids in experimental animals with ulcerative colitis induced by dextran sodium sulfate (DSS). Both butyric and isobutyric acids can improve the health of the intestinal microbiome.

Moreover, consumption of taxifolin has been found to significantly increase the expression of GPR41 and GPR43 receptors, which are important mediators in the interaction between the human host and the gut microbiome (Ang & Ding, 2016).

According to recent reports, Taxifolin has been found to inhibit the expression of TNF-α, IL-1β, and IL-6 in the colonic tissue. It also helps protect advanced glycation end products and mitochondrial damage and aids in the elimination of amyloid beta protein. This is necessary for preventing cerebral myeloid angiopathy, characterized by the accumulation of this protein in the walls of cerebral vessels, leading to hemorrhages and microinfarcts. Taxifolin also has a notable effect on antioxidant activity (we recommend reviewing Chapter 9: The Role of Phenolic Compounds in the Antioxidant Activity of *Cicer arietinum*).

### 3.2. Isoflavones

Isoflavones belong to the group of phenolic compounds. In legumes, the most frequently reported are: biochanin A (5,7-dihydroxyflavone-4′-methoxyflavone, calicosin (7,3′-dihydroxy-4′- methoxyisoflavone, formononetin (7-hydroxy-4′-methoxyisoflavone, formononetin (7-hydroxy-4′-methoxyisoflavone), genistein (5,7,4′-trihydroxyisoflavone, trifolirhizine (maackiain-3-O-β-d-glucopyranoside, ononin (7-O-β- d-glucosyl-7-hydroxy-4'-methoxyisoflavone, sisotrin (7-O-β- d -glucosyl-5,7-dihydroxy-4′-methoxyisoflavone, (Zhao et al., 2009). These molecules have the particularity of having a structure similar to estrogens in humans; for this reason, they are also known as phytoestrogens. It has been proposed that the intestinal microbiota plays a crucial role in metabolizing conjugated isoflavones by hydrolyzing them and releasing the aglycone form, which is more readily absorbed in the small intestine. This process enhances the absorption, metabolism, and reconjugation of isoflavones, contributing to their bioavailability and effectiveness in the body. Among the best-known biological activities for isoflavones are those related to the menstrual cycle associated with the climacteric in premenopausal women, prolonging the cycle; in women in postmenopausal stages, there is a reduction of 40 to 55% of the characteristic symptoms (Sarao et al., 2022).

Other biological activities attributed to isoflavones are antioxidant, anticancer, estrogenic, insecticidal, antifungal, antimicrobial, and contraceptive properties (HaiRong et al., 2012; Wei et al., 2017; Domínguez-Arispuro et al., 2021). Among the mechanisms of action reported for the activity of these compounds is competition with estradiol for binding to estrogen receptors and stimulation of the expression of specific markers of estrogen activity (p52) (Januario et al., 2005). Both genistein and other phytoestrogens have also been postulated to inhibit tumor cell growth by intervening in the tyrosine kinase activity of activated growth factor receptors and cytoplasmic tyrosine kinase and by inhibiting DNA topoisomerases, which are essential for mitogenic signal transduction (Turner et al., 2003).

### 3.3. Saponins

Saponins are molecules of a carbohydrate linked to a triterpenoid or sterol, characterized by their ability to form stable foams in an aqueous medium similar to soap in aqueous solutions. Kerem et al. (2005) reported that the chickpea seed contains mainly type B saponins bound to 2,3-dihydro-2,5-dihydroxy-6-methyl-4 H -pyran-4-one (DDMP). This glycoside is sensitive to heat; however, due to its chemical structure, it can confer antioxidant activity.

Among the main biological activities attributed to saponins, reduced cholesterol absorption has been evidenced by reducing its plasmatic concentration in humans and animals and protecting against oxidative damage (Conh et al., 2010). Other benefits are inhibition of dental caries and platelet aggregation. Besides treating hypercalciuria in humans, they are useful as an antidote to acute lead poisoning (Kerem et al., 2005; Serventi et al., 2013).

### 3.4. Alpha Galactosides

The alpha-galactosides or oligosaccharides identified in chickpea seeds are raffinose, stachyose, verbascose, galactopyranosylciceritol, ciceritol, and galactopinitol, which are considered anti-nutritional factors and due to the lack of alpha-galactosidases in humans their consumption produces flatulence (Llorach et al., 2019; Arunraj et al., 2020). However, these compounds can be eliminated by soaking the seed because they are water-soluble; they are also sensitive to temperature, so they can also be reduced with heat treatments (cooking, extrusion, and roasting), fermentation, or during germination as, in this last process, the enzyme α-galactosidase is activated to release glucose molecules essential for the development of the seedling. In humans, it has been proposed that these compounds are capable of increasing the synthesis of short-chain fatty acids (butyrate and propionate) by the intestinal microbiota (Thirunathan and Manickavasagan, 2019; Arunraj et al., 2020).

## 3.5. Phytates

Phytates, also known as inositol phosphates, are the main form of phosphorus storage in chickpeas, which is not bioavailable to humans due to the lack of phytases, the enzymes responsible for releasing phosphorus. Phytates are also considered anti-nutritional due to the chelating complexes they form with metallic cations of Ca, Fe, K, Mg, Mn, and Zn, making them insoluble and unavailable, and the presence of proteins inhibiting their absorption (Gupta et al., 2015; Bohn et al., 2008).

In contrast, it has been reported that they can also have beneficial health effects; they lower inositol phosphates, such as IP4 and IP3, they may play a beneficial role in mediating cellular responses and have been shown to play a role in second messenger transduction systems (Zajdel et al., 2013). On the other hand, larger molecules such as myo-inositol-hexakisphosphate (IP6) can help treat diseases like diabetes. In this sense, it has been shown that food supplements rich in inositol improve glycemic indices and regulate the activities of some metabolic enzymes involved in lipid and carbohydrate metabolism (Omoruyi et al., 2020). In experimental animals, they can reduce serum glucose and leptin, as well as the rate of insulin resistance in a model of streptozotocin-induced diabetes (Foster et al., 2016). Another proposed mechanism regarding the antidiabetic effect is the decrease in starch digestibility, related to the ability of phytates to bind to minerals. In this sense, when chelating calcium, the enzyme alpha-amylase necessary for starch hydrolysis is inactivated, thereby influencing the serum glucose concentration (Lee et al. 2006).

Phytates also exhibit antioxidant activity by inhibiting the formation of hydroxyl radicals, which are catalyzed by iron (Bohn et al., 2008). Furthermore, research has shown that phytates can inhibit the crystallization of calcium salts, potentially preventing the formation of kidney stones (Grases and Costa-Bauza, 2019).

## Conclusion

Throughout this chapter, it has been demonstrated that both in vivo and in vitro models show considerable evidence of the health benefits of the bioactive compounds present in chickpea seeds. These compounds have significant antioxidant activity and can be used as a natural source of antioxidants. Moreover, they have the potential to stimulate the immune system and reduce various types of inflammation in the human body. However, most of these studies use isolated, concentrated, and purified bioactive compounds. This process can be time-consuming and expensive and can lead to low extraction yields and losses. Therefore, further studies are required to investigate the consumption of whole seeds, as bioactive compounds in the seeds might act synergistically or antagonistically. The effect is produced after different processes in the seed, which can be as simple as toasting, cooking, sprouting, fermentation, or being incorporated into everyday dishes or more sophisticated dishes like hummus. Additionally, the mechanisms of action involved still need to be demonstrated to understand the biological effect of whole seeds. Furthermore, nutrigenetic and nutrigenomic studies could complement the research to see the impact of diets rich in legumes, particularly chickpeas, on human health.

## References

Acevedo-Martínez, K. A., Yang, M. M., & González de Mejia, E. (2021). Technological properties of chickpea (*Cicer arietinum*): Production of snacks and health benefits related to type-2 diabetes. *Comprehensive Reviews in Food Science and Food Safety,* 20(4), 3762-3787. doi: 10.1111/1541-4337.12762.

Alam, W., Khan, H., Shah, M. A., Cauli, O., & Saso, L. (2020). Kaempferol as a dietary anti-inflammatory agent: current therapeutic standing. *Molecules*, *25*(18), 4073. doi: 10.3390/molecules25184073.

Ang, Z., & Ding, J. L. (2016). GPR41 and GPR43 in obesity and inflammation–protective or causative?. *Frontiers in immunology*. doi: 10.3389/fimmu.2016.00028.

Arevalo, I., Guzmán-Maldonado, S. H., Sanchez, S. M. M., Acosta-Gallegos, J. A. Steaming and Toasting Reduce the Nutrimental Quality, Total Phenols and Antioxidant Capacity of Fresh Kabuli Chickpea (*Cicer arietinum L.*). *Plant Foods for Human Nutrition.* 75(4):628-634. doi: 10.1007/s11130-020-00857-5.

Arunraj, R., Skori, L., Kumar, A., Hickerson, N. M., Shoma, N., & Samuel, M. A. (2020). Spatial regulation of alpha-galactosidase activity and its influence on raffinose family oligosaccharides during seed maturation and germination in *Cicer arietinum. Plant signaling & behavior,* 15(8), 1709707. doi: 10.1080/15592324.2019.1709707.

Berhe, M., Subramanyam, B., Demissie, G., Chichaybelu, M., Abera, F. A., Mahroof, R., & Harvey, J. (2024). Impact of storage technologies and duration on insect pest population, post-harvest losses, and seed quality of stored chickpea in Ethiopia. *Pest Management Science*, *80*(2), 518-532. doi: 10.1002/ps.7781.

Bohn, L., Meyer, A. S., & Rasmussen, S. K. (2008). Phytate: impact on environment and human nutrition. A challenge for molecular breeding. *Journal of Zhejiang University Science B*, *9*(3), 165-191. doi: 10.1631/jzus.B0710640.

Campos-Vega, R., Oomah, B. D., Loarca-Piña, G., & Vergara-Castañeda, H. A. (2013). Common beans and their non-digestible fraction: cancer inhibitory activity—an overview. *Foods*, 2(3), 374-392. doi:10.3390/foods2030374.

Cardador-Martínez, A., Martínez-Tequitlalpan, Y., Gallardo-Velazquez, T., Sánchez-Chino, X. M., Martínez-Herrera, J., Corzo-Ríos, L. J., & Jiménez-Martínez, C. (2020). Effect of Instant Controlled Pressure-Drop on the Non-Nutritional Compounds of Seeds and Sprouts of Common Black Bean (*Phaseolus vulgaris L.*). *Molecules.* 25 (6). doi: 10.3390/molecules25061464.

Chandrasekaran, S., Luna-Vital, D., & de Mejia, E. G. (2020). Identification and comparison of peptides from chickpea protein hydrolysates using either bromelain or gastrointestinal enzymes and their relationship with markers of type 2 diabetes and bitter *Nutrients*, *12*(12), 3843. doi: 10.3390/nu12123843.

Chávez-Ontiveros, J., Reyes-Moreno, C., Ramírez-Torres, G. I., Figueroa-Salcido, O. G., Arámburo-Gálvez, J. G., Montoya-Rodríguez, A., Ontiveros, N. & Cuevas-Rodríguez, E. O. (2022). Extrusion Improves the Antihypertensive Potential of a Kabuli Chickpea (*Cicer arietinum L.*) Protein Hydrolysate. *Foods*, *11*(17), 2562. doi: 10.3390/foods11172562.

Cid-Gallegos, M. S., Corzo-Ríos, L. J., Jiménez-Martinez, C., & Sánchez-Chino, X. M. (2022). Protease Inhibitors from Plants as Therapeutic Agents-A Review. *Plant Foods for Human Nutrition,* 77(1), 20-29. doi: 10.1007/s11130-022-00949-4.

Cid-Gallegos, M. S., Sánchez-Chino, X. M., Álvarez-González, I., Madrigal-Bujaidar, E., Vásquez-Garzón, V. R., Baltiérrez-Hoyos, R., Villa-Treviño S., Dávila-Ortiz, G. & Jiménez-Martínez, C. (2020). Modification of in vitro and in vivo antioxidant activity by consumption of cooked chickpea in a colon cancer model. *Nutrients*, *12*(9), 2572. doi: 10.3390/nu12092572.

Cohn J. S., Kamili, A., Wat, E., Chung, R. W., & Tandy, S. (2010). Reduction in intestinal cholesterol absorption by various food components: mechanisms and implications. Atherosclerosis Supplements, 11(1), 45-48. doi: 10.1016/j.atherosclerosissup. 2010.04.004.

Corzo-Ríos, L. J., Sánchez-Chino, X. M., Cardador-Martínez, A., Martínez-Herrera, J., & Jiménez-Martínez, C. (2020). Effect of cooking on nutritional and non-nutritional compounds in two species of Phaseolus (*P. vulgaris* and *P. coccineus*) cultivated in Mexico. *International Journal of Gastronomy and Food Science,* 20, 100206. doi: 10.1016/j.ijgfs.2020.100206.

Corzo-Ríos, L. J., Jiménez-Martínez, C., Cid-Gallegos, M. S., Cardador-Martínez, A., Martínez-Herrera, J., & Sánchez-Chino, X. M. (2022). Chemical and non-nutritional modification of faba bean (*Vicia faba*) due to the effect of roasting and boiling.

*International Journal of Gastronomy and Food Science*, *30*, 100622. doi:10.1016/j.ijgfs.2022.100622.

Costantini, M., Summo, C., Centrone, M., Rybicka, I., D'Agostino, M., Annicchiarico, P., Caponio, F., Pavan, S., Tamma, G., & Pasqualone, A. (2021). Macro- and Micro-Nutrient Composition and Antioxidant Activity of Chickpea and Pea Accessions. *Polish Journal of Food and Nutrition Sciences*, 71(2), 177–185. doi: 10.31883/pjfns/135813.

Di Pasquale M. G. (2009). The essentials of essential fatty acids. *Journal of Dietary Supplements* 6, 143–161. doi: 10.1080/19390210902861841.

Domínguez-Arispuro, D. M., Canizalez-Román, A., Reyes-Moreno, C., Milán-Noris, A. K., Milán-Carrillo, J., Acosta-Smith, E., & Cuevas-Rodríguez, E. O. (2021) Isoflavones from black chickpea (*Cicer arietinum* L) sprouts with antioxidant and antiproliferative activity. *Saudi Journal of Biological Sciences,* 28 (1). doi: 10.1016/j.sjbs.2020.11.048.

Eatemadi, A., Aiyelabegan, H. T., Negahdari, B., Mazlomi, M. A., Daraee, H., Daraee, N., Eatemadi, R. & Sadroddiny, E. (2017). Role of protease and protease inhibitors in cancer pathogenesis and treatment. *Biomedicine & Pharmacotherapy,* 86:221-231. doi: 10.1016/j.biopha.2016.12.021.

Foster, S. R., Omoruyi, F. O., Bustamante, J., Lindo, R. L., & Dilworth, L. L. (2016). The effect of combined inositol hexakisphosphate and inositol supplement in streptozotocin-induced type 2 diabetic rats. *International Journal of Experimental Pathology*, *97*(5), 397-407. doi: 10.1111/iep.12210.

Grases, F., & Costa-Bauza, A. (2019). Key aspects of myo-inositol hexaphosphate (phytate) and pathological calcifications. *Molecules*, *24*(24), 4434. doi: 10.3390/molecules24244434.

Gupta, R. K., Gangoliya, S. S., & Singh, N. K. (2015). Reduction of phytic acid and enhancement of bioavailable micronutrients in food grains. *Journal of food science and technology*, *52*(2), 676-684. doi:10.1007/s13197-013-0978-y.

Han, X., Li, M., Sun, L., Liu, X., Yin, Y., Hao, J., & Zhang, W. (2022). p-hydroxybenzoic acid ameliorates colitis by improving the mucosal barrier in a gut microbiota-dependent manner. *Nutrients,* 14(24), 5383. doi: 10.3390/nu14245383.

HaiRong, M., HuaBo, W., Zhen, C., Yi, Y., ZhengHua, W., Madina, H., ... & Haji Akber, A. (2013). The estrogenic activity of isoflavones extracted from chickpea Cicer arietinum L sprouts in vitro. *Phytotherapy Research*, *27*(8), 1237-1242. doi: 10.1002/ptr.4858.

Harini, S., Adilaxmamma, K., Mohan, E. M., Srilatha, C., & Raj, M. A. (2015). Antihyperlipidemic activity of chickpea sprouts supplementation in ovariectomy-induced dyslipidemia in rats. *Journal of Ayurveda and Integrative Medicine*, *6*(2), 104. doi: 10.4103/0975-9476.146546.

Hellinger, R. & Gruber C. W. (2019) Peptide-based protease inhibitors from plants. *Drug Discovery Today* 24(9):1877–1889. doi: 10.1016/j.drudis.2019.05.026.

Honda, T., Kabashima, K., & Kunisawa, J. (2023). Exploring the roles of prostanoids, leukotriens, and dietary fatty acids in cutaneous inflammatory diseases: Insights from pharmacological and genetic approaches. *Immunological Reviews*, *317*(1), 95-112. doi: 10.1111/imr.13193.

Imran, M., Salehi, B., Sharifi-Rad, J., Aslam Gondal, T., Saeed, F., Imran, A., Shahbaz, M., Tsouh Fokou, P. V., Haroon Khan M. U., Guerreiro S. G., Martins, N. & Estevinho, L. M. (2019). Kaempferol: A key emphasis to its anticancer potential. *Molecules*, *24*(12), 2277. doi: 10.3390/molecules24122277.

Januario, A. H., Lourenco, M. V., Domezio, L. A., Pietro, R. C. L. R., Castilho, M. S., Tomazela, D. M., Fernandes da Silva, M. F. G., Paulo Cezar Vieira, P. C., João Batista Fernandes, J. B. & Suzelei de Castro França S. C. (2005). Isolation and structure determination of bioactive isoflavones from callus culture of Dipteryx odorata. *Chemical and Pharmaceutical Bulletin*, 53(7), 740–742. doi: 10.1248/cpb.53.740.

Jukanti A. K., Gaur P. M., Gowda C. L. L., Chibbar R. N. (2012). Nutritional quality and health benefits of chickpea (*Cicer arietinum* L.): A review. *British Journal of Nutrition,* 108(Suppl. 1), S11–S26. doi: 10.1017/S0007114512000797.

Kaur R., Prasad K. (2021). Technological, processing and nutritional aspects of chickpea (*Cicer arietinum*) – A review. *Trends in Food Science & Technology.* 109, 448–463. doi: 10.1016/j.tifs.2021.01.044.

Kerem Z, German-Shashoua H, Yarden O. Microwave-assisted extraction of bioactive saponins from chickpea (*Cicer arietinum* L.) *Journal of the Science of Food and Agriculture* 2005; 85:406–412. doi: 10.1002/jsfa.1989.

Khurana, V., & Goswami, B. (2022). Angiotensin converting enzyme (ACE). *Clinica Chimica Acta*, *524*, 113-122. doi: 10.1016/j.cca.2021.10.029.

Lee, S.-H., Park, H.-J., Chun, H.-K., Cho, S.-Y., Cho, S.-M., & Lillehoj, H. S. (2006). Dietary phytic acid lowers the blood glucose level in diabetic KK mice. *Nutrition Research*, *26*(9), 474-479. doi:10.1016/j.nutres.2006.06.017.

León-López, L., Escobar-Zúñiga, Y., Salazar-Salas, N. Y., Mora Rochín, S., Cuevas-Rodríguez, E. O., Reyes-Moreno, C., & Milán-Carrillo, J. (2020). Improving polyphenolic compounds: Antioxidant activity in chickpea sprouts through elicitation with hydrogen peroxide. *Foods*, 9(12), 1791. doi: 10.3390/foods9121791.

Lorenzo, J. M., Munekata, P. E., Gomez, B., Barba, F. J., Mora, L., Perez-Santaescolastica, C., & Toldra, F. (2018). Bioactive peptides as natural antioxidants in food products–A review. *Trends in Food Science & Technology, 79*, 136-147. doi: 10.1016/j.tifs.2018.07.003.

Li, W., Zhang, L., Xu, Q., Yang, W., Zhao, J., Ren, Y., Yu, Z. & Ma, L. (2022). Taxifolin alleviates DSS-induced ulcerative colitis by acting on gut microbiome to produce butyric Acid. *Nutrients*, *14*(5), 1069. doi: 10.3390/nu14051069.

Llorach, R., Favari, C., Alonso, D., Garcia-Aloy, M., Andres-Lacueva, C., & Urpi-Sarda, M. (2019). Comparative metabolite fingerprinting of legumes using LC-MS-based untargeted metabolomics. *Food Research International,* 126, 108666. doi: 10.1016/j.foodres.2019.108666.

Madurapperumage, A., Tang, L., Thavarajah, P., Bridges, W., Shipe, E., Vandemark, G., & Thavarajah, D. (2021). Chickpea (*Cicer arietinum L.*) as a source of essential fatty acids–a biofortification approach. *Frontiers in Plant Science*, *12*, 734980. doi: 10.3389/fpls.2021.734980.

Magalhães, S. C. Q., Taveira, M., Cabrita, A. R. J., Fonseca, A. J. M., Valentão, P., & Andrade, P. B. (2017). European marketable grain legume seeds: Further insight into phenolic compounds profiles. *Food Chemistry*, 215, 177–184. doi: 10.1016/j.foodchem.2016.07.152.

Magee, P. J., Owusu-Apenten, R., McCann, M. J., Gill, C. I., & Rowland, I. R. (2012). Chickpea (*Cicer arietinum*) and other plant-derived protease inhibitor concentrates inhibit breast and prostate cancer cell proliferation *in vitro. Nutrition and cancer*, 64(5), 741-748. doi: 10.1080/01635581.2012.688914.

Mathesius, U. (2022). Are legumes different? Origins and consequences of evolving nitrogen fixing symbioses. *Journal of Plant Physiology*, 276, 153765. doi: 10.1016/j.jplph.2022.153765.

Medina-Godoy, S., Ambriz-Pérez, D. L., Fuentes-Gutiérrez, C. I., Germán-Báez, L. J., Gutiérrez-Dorado, R., Reyes-Moreno, C., & Valdez-Ortiz, A. (2012). Angiotensin-converting enzyme inhibitory and antioxidative activities and functional characterization of protein hydrolysates of hard-to-cook chickpeas. *Journal of the Science of Food and Agriculture*, 92(9), 1974-1981. doi: 10.1002/jsfa.5570.

Megías, C., Pedroche, J., Yust, M. M., Girón-Calle, J., Alaiz, M., Millán, F., & Vioque, J. (2007). Affinity purification of copper chelating peptides from chickpea protein hydrolysates. *Journal of Agricultural and Food Chemistry,* 55(10), 3949-3954. doi: 10.1021/jf063401s.

Milán-Noris, A. K., Gutierrez-Uribe, J. A., & Serna-Saldivar, S. O. (2023). Influence of soaking and boiling on flavonoids and saponins of nine desi chickpea cultivars with potential antiproliferative effects. *Journal of Food Measurement and Characterization.* doi: 10.1007/s11694-023-01861-4.

Muzquiz, M., Varela, A., Burbano, C., Cuadrado, C., Guillamón, E., & Pedrosa, M. M. (2012). Bioactive compounds in legumes: pronutritive and antinutritive actions. Implications for nutrition and health. *Phytochemistry Reviews*, 11(2), 227–244. doi:10.1007/s11101-012-9233-9.

Ogawa A., Suzuki Y., Aoyama T., Takeuchi H. (2009). Dietary alpha-linolenic acid inhibits angiotensin-converting enzyme activity and mrna expression levels in the aorta of spontaneously hypertensive rats. *Journal of Oleo Science.* 58, 355–360. doi: 10.5650/jos.58.355.

Omoruyi, F. O., Stennett, D., Foster, S., & Dilworth, L. (2020). New frontiers for the use of IP6 and inositol combination in treating diabetes mellitus: A review. *Molecules*, *25*(7), 1720. doi: 10.3390/molecules25071720.

Pasqualone, A., Summo, C., De Angelis, D., Cucci, G., Caranfa, D., & Lacolla, G. (2021). Effect of mineral and organic fertilization on desi and kabuli chickpea (*Cicer arietinum L.*): Plant growth and production, hydration properties, bioactive compounds, and antioxidant activity. *Plants,* 10(7), 1441. doi: 10.3390/ plants10071441.

Primi, R., Ruggeri, R., Ronchi, B., Bernabucci, U., Rossini, F., Martin-Pedrosa, M., & Danieli, P. P. (2019). Sowing date and seeding rate affect bioactive compound contents of chickpea grains. *Animals*, 9(8), 571. doi: 10.3390/ani9080571.

Quintero-Hilario, C. C., Esparza-Torres, F., García-Mateos M. R., Ybarra-Moncada, M. C., & Hernández-Ramos, L. (2019). Effect of roasting on Maya nut's nutritional value and antioxidant components (*Brosimum alicastrum*: Moraceae) Ef. *Revista Chapingo Serie Horticultura,* 25(3). doi:10.5154/r.rchsh.2019.03.007.

Quintero-Soto, M. F., Chávez-Ontiveros, J., Garzón-Tiznado, J. A., Salazar-Salas, N. Y., Pineda-Hidalgo, K. V., Delgado-Vargas, F., & López-Valenzuela, J. A. (2021). Characterization of peptides with antioxidant activity and antidiabetic potential obtained from chickpea (*Cicer arietinum* L.) protein hydrolyzates. *Journal of Food Science*, *86*(7), 2962-2977. doi:10.1111/1750-3841.15778.

Rachwa-rosiak, D., & Nebesny, E. (2015). Chickpeas — Composition, Nutritional Value, Health Benefits, Application to Bread and Snacks: A Review. *Critical Reviews in Food Science and Nutrition* 55(8):1137-1145. doi:10.1080/10408398.2012.687418.

Real Hernández, L. M., & González de Mejia, E. (2019). Enzymatic production, bioactivity, and bitterness of chickpea (*Cicer arietinum*) peptides. Comprehensive *Reviews in Food Science and Food Safety,* 18(6), 1913–1946. doi: 10.1111/1541-4337.12504.

Roberfroid M., Gibson G. R., Hoyles L., McCartney A. L., Rastall R., Rowland I., Wolvers, D., WatzL, B., Szajewska, H., Stahl, B., Guarner, F., Respondek, F., Whelan, K., Coxam, V., Davicco, M., Léotoing, L., Wittrant, Y., Delzenne, N. M., Cani, P. D., Neyrinck A. M., Meheus A. (2009) Prebiotic effects: metabolic and health benefits. *British Journal of Nutrition*101, 1–63. doi: 10.1017/S0007114510003363.

Saito, S., Tanaka, M., Satoh-Asahara, N., Carare, R. O., & Ihara, M. (2021). Taxifolin: a potential therapeutic agent for cerebral amyloid angiopathy. *Frontiers in pharmacology,* 12, 643357. doi: 10.3389/fphar.2021.643357.

Saleh A. Alajaji & Tarek A. El-Adawy. (2006). Nutritional conposition of chickpea (*Cicer arietinum* L.) as affected by microwave cooking and other traditional cooking methods. *Journal of Food Composition and Analysis*, 19(8), 8006-812. doi: 10.1016/j.jfca.2006.03.015.

Sáez, G. D., Sabater, C., Fara, A., & Zárate, G. (2022). Fermentation of chickpea flour with selected lactic acid bacteria for improving its nutritional and functional properties. *Journal of Applied Microbiology*, *133*(1), 181-199. doi: 10.1111/jam.15401.

Sánchez-Chino, X. M., Jiménez-Martínez, J., Vásquez-Garzón, V. R., Álvarez-González, I., Villa-Treviño, S., Madrigal-Bujaidar, E., Dávila-Ortiz, G. & Baltierrez-Hoyos, B. (2017). Cooked chickpea consumption inhibits colon carcinogenesis in mice induced with azoxymethane and dextran sulfate sodium. J*ournal of the American College of Nutrition,* 36(5), 391-398. doi: 10.1080/07315724.2017.1297744.

Sarao, L., Kaur, S., Malik, T., & Singh, A. (2022). Chapter 19 - Genistein and daidzein. In J. Kour & G. A. Nayik (Eds.), *Nutraceuticals and Health Care* (pp. 331-341). Academic Press. doi: 10.1016/B978-0-323-89779-2.00016-8.

Serventi, L., Chitchumroonchokchai, C., Riedl, K. M., Kerem, Z., Berhow, M. A., Vodovotz, Y., Schwartz, S.J & Failla, M. L. (2013). Saponins from soy and chickpea: Stability during beadmaking and in vitro bioaccessibility. *Journal of agricultural and food chemistry*, *61*(27), 6703-6710. doi: 10.1021/jf401597y.

Singh M. (2005). Essential fatty acids, DHA human brain. *Indian Journal of Pediatrics,* 72, 239–242. doi: 10.1007/BF02859265.

Sofi, S. A., Rafiq, S., Singh, J., Mir, S. A., Sharma, S., Bakshi, P., McClements, D. J., Mousavi Khaneghah, A., & Dar, B. N. (2023). Impact of germination on structural, physicochemical, techno-functional, and digestion properties of desi chickpea (*Cicer arietinum* L.) flour. *Food Chemistry*, 405, 135011. doi: 10.1016/j.foodchem. 2022.135011.

Tas, A. A., & Shah, A. U. (2021). The replacement of cereals by legumes in extruded snack foods: Science, technology and challenges. *Trends in Food Science and Technology*, 116(April), 701–711. doi: 10.1016/j.tifs.2021.08.016.

Thirunathan, P., & Manickavasagan, A. (2019). Processing methods for reducing alpha-galactosides in pulses. *Critical Reviews in Food Science and Nutrition*, 59(20), 3334-3348. doi: 10.1080/10408398.2018.1490886.

Turner, N. J., Thomson, B. M., & Shaw, I. C. (2003). Bioactive isoflavones in functional foods: the importance of gut microflora on bioavailability. *Nutrition Reviews,* 61(6), 204-213. doi: 10.1301/nr.2003.jun.204-213.

Ulug, S. K., Jahandideh, F., & Wu, J. (2021). Novel technologies for the production of bioactive peptides. *Trends in Food Science & Technology, 108*, 27-39. doi: 10.1016/j.tifs.2020.12.002.

Vidal-Valverde, C., Frías, J., & Valverde, S. (1993). Changes in the carbohydrate composition of legumes after soaking and cooking. *Journal of the American Dietetic Association*, 93(5), 547-550. doi: 10.1016/0002-8223(93)91814-7.

Wang, J., Li, Y., Li, A., Liu, R. H., Gao, X., Li, D., Kou, X., & Xue, Z. (2021). Nutritional constituent and health benefits of chickpea (*Cicer arietinum* L.): A review. *Food Research International*, 150, 110790. doi:10.1016/j.foodres.2021.110790.

Wallace, T. C., Murray, R., & Zelman, K. M. (2016). The nutritional value and health benefits of chickpeas and hummus. *Nutrients*, *8*(12), 766. doi: 10.3390/nu8120766.

Wei, Y., Li, P., Li, B., Gao, J., Wang, D., Qin, L., ... & Liu, T. (2017). Study of the Hypoglycemic Activity of Derivatives of Isoflavones from *Cicer arietinum* L. *Evidence-Based Complementary and Alternative Medicine*, 2017. doi: 10.1155/2017/8746823.

Wood, J. A. and Grusak, M. A. (2007) Nutritional value of chickpea. In: Yadav, S. S., Redden, B., Chen, W. and Shar- ma, B., Eds., Chickpea Breeding and Management, CAB International, Wallingford, 101-142. doi:10.1079/9781845932138.005.

Wood, J. A., Knights, E. J., & Choct, M. (2011). Morphology of chickpea seeds (*Cicer arietinum* L.): comparison of desi and kabuli types. *International Journal of Plant Sciences*, 172(5), 632-643. DOI. 10.1086/659456.

Xiao, S., Li, Z., Zhou, K., & Fu, Y. (2023). Chemical composition of kabuli and desi chickpea (*Cicer arietinum* L.) cultivars grown in Xinjiang, China. *Food Science & Nutrition*, 11(1), 236–248. doi: 10.1002/fsn3.3056.

Yust, M. D. M., Millán-Linares, M. D. C., Alcaide-Hidalgo, J. M., Millán, F., & Pedroche, J. (2012). Hypocholesterolaemic and antioxidant activities of chickpea (*Cicer arietinum L.*) protein hydrolysates. *Journal of the Science of Food and Agriculture*, *92*(9), 1994-2001. doi: 10.1002/jsfa.5573.

Zajdel, A., Wilczok, A., Węglarz, L., & Dzierżewicz, Z. (2013). Phytic acid inhibits lipid peroxidation in vitro. *BioMed research international, 2013.* doi: 10.1155/2013/147307.

Zhao, S., Zhang, L., Gao, P., & Shao, Z. (2009). Isolation and characterisation of the isoflavones from sprouted chickpea seeds. *Food Chemistry*, 114(3), 869-873. DOI. 10.1016/j.foodchem.2008.10.026.

Zia-Ul-Haq M., Ahmad M., Iqbal S., Ahmad S., Ali H. (2007). Characterization and Compositional Studies of Oil from Seeds of Desi Chickpea (*Cicer arietinum* L.) Cultivars Grown in Pakistan. *Journal of the American Oil Chemists' Society* 84, 1143–1148. doi: 10.1007/s11746-007-1136-3.

## Chapter 4

# *Glycine max:* Secondary Metabolites

**A. Ortiz-Fernández**
**M. A. A. Dzul-Cervantes**
**E. Pérez-Pacheco**
**J. C. Canto-Pinto**
**R. J. Estrada-León**
**and Y. A. Chim-Chi***
Tecnológico Nacional de México, Campus Calkiní, Calkiní, Campeche, México

### Abstract

Soybean, scientifically known as *Glycine max* , is a significant crop worldwide due to its high nutritional value. This legume contains essential nutrients such as proteins, amino acids (including lysine), unsaturated fatty acids (linoleic and linolenic acid), dietary fiber, vitamins, and minerals. However, the researchers are particularly interested in its secondary metabolites and potential health benefits. Secondary metabolites are molecules synthesized by plants that are not essential for their survival. These metabolites are influenced by environmental factors such as soil, rain, and temperature, as well as the characteristics of the plant itself. In the case of *Glycine max*, many researchers have reported the presence of several secondary metabolites, including lectins, saponins, phytates, isoflavones, and phenolic compounds. Lectins, specifically agglutinins, are crucial in separating stem cells from human bone marrow. On the other hand, Saponins exhibit antimicrobial properties and can effectively combat bacteria, fungi, and

* Corresponding Author's Email: yachim@itescam.edu.mx

In: Legumes: Beyond Their Nutritional Value
Editors: C. Jiménez-Martínez, M. F. Juárez-Chairez and M. S. Cid-Gallegos
ISBN: 979-8-89530-440-2

insects. Additionally, they can help reduce cholesterol levels in the blood. Although phytates are considered antinutrients, they also have positive effects, such as anti-inflammatory and antioxidant properties and the ability to prevent certain types of cancer. Isoflavones have a phytoestrogen effect and are beneficial to human health. They can help prevent cancer, atherosclerosis, and cardiovascular diseases. Lastly, phenolic compounds are antioxidants, the most important ones being chlorogenic acid, caffeic acid, ferulic acid, and coumaric acid. Therefore, this chapter will discuss the secondary metabolites found in soybeans, their effects, and their potential applications in biotechnology and human health.

**Keywords**: *Glycine max*, secondary metabolites, antioxidant, human health

## 1. Introduction

The soybean belongs to *Rosaceae* order, *Leguminosae* or *Papillonaceae* or Fabaceae, subfamily Papilionoidae, the genus Glycine, and the cultivar *Glycine max*. The soybean originated in China, probably in the north and central regions, 4,000-5,000 years ago. Soybean was introduced in Europe around 1712 by a German botanist, Engelbert Kaempfer. Later, Carl von Linné gave soybeans a genetic name, *Glycine max*. However, due to the climate and soil conditions, soybean production has been limited in Europe. Soybean is a singular food because of its rich nutrient content, which contains protein (35-45%), oligosaccharides, dietary fiber, phytochemicals (especially isoflavones), and minerals (Scheppach et al., 2004). In 1999, the Food and Drug Administration (FDA) approved a health claim based on the role of soybean protein in reducing the risk of coronary heart disease. Research has shown that frequent soybean protein consumption lowers cholesterol levels. Results from a recent study suggest that soybean dietary fiber plays a role in reducing cholesterol levels in some hyperlipidemic individuals and has a major protective effect on cardiovascular disease. Moreover, it improves glucose tolerance in some diabetic patients; it increases the wet fecal weight and reduces caloric density in some foods (Mateos-Aparicio et al., 2008). This chapter will describe secondary soybean metabolites, their effects on human health, and their applications in biotechnology.

## 2. Secondary Metabolites in Soybean

Secondary metabolites are molecules synthesized by plants that are not essential for survival. *Glycine max* produces a variety of these metabolites, including lectins, saponins, phytates, isoflavones, and phenolic compounds (Graham et al., 2008; Lee et al., 2014). The effects of these molecules are outlined below.

### 2.1. Lectins

Lectins are proteins that play crucial roles in various plant processes, including defense and stress signaling. They possess at least one non-catalytic domain, which allows them to bind to specific glycan structures reversibly. Around 359 potential lectin genes have been identified in the soybean genome, corresponding to homologs of nine out of twelve known lectin families (Van Holle & Van Damme, 2015).

Lectins can agglutinate cells, which is a well-known physiological effect. This effect depends on their specific, high-affinity binding to particular carbohydrate moieties on the cell surface. Soybean lectins are glycoproteins that have a tetrameric structure and weigh 120 kD. They make up 1-2% of seed protein mass. However, adverse nutritional effects have been observed after consuming raw soybean meal. These effects have been attributed to endogenous inhibitors of digestive enzymes, poor digestibility, and lectins (De Mejia et al., 2003).

In rats, soybean lectins can stimulate the release of cholecystokinin, which stimulates pancreatic enzyme secretion. Lectin activity persists during the passage through the gastrointestinal tract. Soybean lectins that contain N-acetyl-D-galactosamine tend to stimulate intestinal cells and can interfere with the absorption of nutrients in the intestine. For these reasons, soybean-derived lectins are traditionally recognized as having antinutritional properties. However, heat treatment can inactivate lectins (Vasconcelos et al., 1997).

Several studies have indicated that plant lectins possess antitumor and anticarcinogenic properties, which could benefit cancer treatment. The pattern of lectin binding sites can suggest the degree of tumor malignancy, consistent with the clinical behavior of cell carcinomas (Ueno et al., 2000). Due to their properties, lectins could serve as versatile biomarkers and have been used in various studies involving histochemical, biochemical, and functional techniques for characterizing cancer cells (Muñoz et al., 2001). Soybean agglutinin is considered the most important lectin found in soybean (De Mejia et al., 2003).

Soybean agglutinin (SBA) is a protein found in soybeans unrelated to fiber. It is also the main antinutritional factor present in the seed. The molecular weight of SBA is around 120 kDa. This protein has a strong binding affinity for N-glyphthalide-d-galactosamine or galactose, and it has a covalently linked oligosaccharide chain. SBA can negatively affect the intestinal health of animals by influencing the structure of the intestine, the barrier function, the mucosal immune system, and the balance of the gastrointestinal microbiota (Pan et al., 2018). However, SBA has positive effects, such as antitumor, antifungal, antiviral, and antibacterial activities (Lin et al., 2008). Breeze et al. (2015) developed a precise and accurate method to measure soybean agglutinin using an enzyme-linked-lectin-assay (ELLA). Validation experiments demonstrated that the ELLA method can reliably recover soybean agglutinin from seeds, and the presence of other components in soybean seeds does not interfere with the analysis. These findings suggest that the ELLA method is both accurate and precise for determining lectin levels in soybean seeds.

Moreover, Tang et al. (2006) conducted an experiment to investigate the effects of purified soybean agglutinin on the growth and immune function of rats. The study involved feeding thirty male Sprague-Dawley rats casein and corn starch-based diets containing 0, 0.05, 0.10, 0.15, or 0.20% soybean agglutinin (w/w) for 20 days. The results showed that purified soybean agglutinin in rat diets negatively affected growth and immune function, including cell-mediated and humoral immune function. However, the mechanism through which soybean agglutinin acts needs to be clarified, and it affects different animals, such as rats, differently. Therefore, further studies are needed to determine how soybean agglutinin suppresses growth and immune function in other animals.

## 2.2. Saponins

Saponins are a diverse group of compounds that get their name from their ability to create stable, soap-like foams in water. Chemically, saponins comprise a carbohydrate attached to a triterpenoid or steroid (Figure 1). These molecules are phytochemicals that taste bitter and may affect health, although the effects are still being debated. Some saponins are found in soybeans and are known as soyasaponins. Soyasaponins are amphiphilic compounds with polar sugar moieties attached to a nonpolar pentacyclic ring. They are divided into two groups, A and B, based on the glycosylation positions of soyasapogenol A and soyasapogenol B. Soyasaponin Aa and soyasaponin Ab are glycosylated at the C-3 and C-22 positions of soyasapogenol A (group A). In contrast, soyasaponin Ba and soyasaponin Bb are glycosylated at the C-3 position of soyasapogenol B (group B). Group A soyasaponins cause a more unpleasant bitter taste in soybean foods than group B saponins (Gou & Chen, 2018).

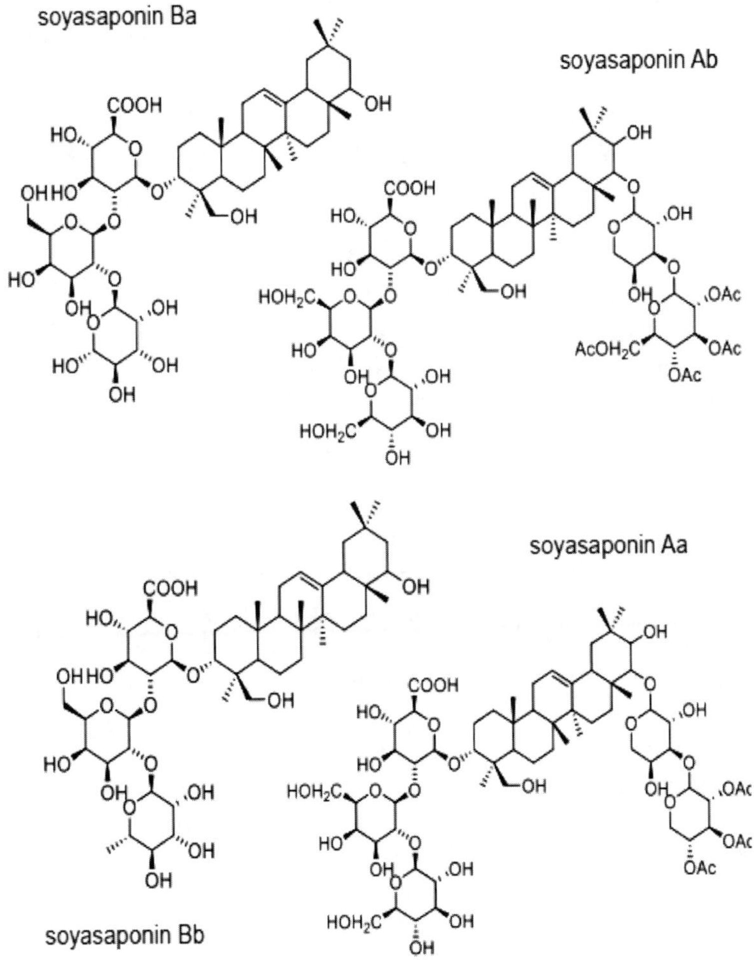


**Figure 1.** Chemical structures of soyasaponins (Bljahhina et al., 2023).

In clinical research, it has been suggested that certain health-promoting components, such as saponins, can positively impact the human body. Saponins can affect the immune system, help protect against cancer, and lower cholesterol levels. Moreover, can help decrease blood lipids and lower blood glucose response. A diet high in saponins can inhibit dental caries and platelet aggregation, treat hypercalciuria in humans, and even act as an antidote against acute lead poisoning.

In epidemiological studies, saponins have been found to have a relationship with a lower incidence of renal stones (Shi et al., 2004). However, the most significant effect of saponins is against cancer. Soybean saponins can induce apoptosis (programmed cell death) in cancer cells, inhibit tumor cell proliferation, and suppress angiogenesis (growth of new blood vessels that nourish tumors). They can also prevent or slow the progression of various types of cancer, such as breast, lung, prostate, and colon cancers (Hwang, 2017; Kerwin, 2004). Moreover, saponins can inhibit the production of pro-inflammatory cytokines and enzymes, thereby reducing inflammation and alleviating symptoms in conditions like arthritis, inflammatory bowel disease, and other inflammatory disorders (Timilsenaet al., 2023).

Finally, saponins have been found to display antimicrobial properties against a broad spectrum of pathogens, such as bacteria, viruses, fungi, and protozoa. They achieve this by disrupting microbial cell membranes and interfering with their replication. As a result, they can potentially be used to develop new antimicrobial agents and improve existing treatments (Dong et al., 2019).

Jun et al. (2002) conducted a study to determine the impact of soybean saponins on AFB1-induced mutagenicity and AFB1-DNA adduct formation. They used *Salmonella typhimurium* and human liver hepatoma (HepG2) cells to measure the effects. Prior research had shown that antioxidants such as L-ascorbic acid, α-tocopherol, and butylated hydroxytoluene (BHT) had antimutagenic properties. The study found that BHT had the strongest antimutagenic activity, followed by saponins, α-tocopherol, and L-ascorbic acid. Soybean saponins significantly reduced the mutagenicity of AFB1 by 52, 64, and 81% at concentrations of 600, 900, and 1,200 μg per plate, respectively. This suggests that soybean saponins may have the ability to block the initiation stage of carcinogenesis (Jun et al., 2002).

Hassan et al. (2009) conducted a study to evaluate the antimicrobial properties of soybean saponins. They tested the saponins against three different types of bacteria: *Staphylococcus aureus* (a gram-positive bacterium), *Salmonella typhimurium*, and *Escherichia coli* (both gram-negative bacteria). They dissolved 25 mg of saponin in 1 mL of tryptic soy broth to create a stock solution of saponin. The study results showed that the saponins effectively inhibited the growth of these bacteria.

## 2.3. Phytates

In soybean seeds, phytates can represent 2-2.5% of the total weight. Due to their chemical structure, these molecules can actively bind to minerals like iron, zinc, and calcium. The formation and stability of phytate complexes in solution depend on various factors, such as pH, ionic strength, supporting electrolyte, temperature, nature, and metal concentrations. Chelation occurs more effectively at medium to higher pH levels, as higher pH levels facilitate deprotonation and increase electronegativity (Wang & Gou, 2021). Phytates in soybeans are known to cause mineral deficiency in people who have grain-based diets, especially in vulnerable groups like infants, teenagers, pregnant women, and the elderly. However, phytates also have health benefits, like reducing cholesterol levels, acting as antioxidants, and possibly being anti-cancer agents (Karkle & Beleia, 2010).

Soybeans contain a molecule called phytic acid, which has a strong attraction to zinc, iron, and trace minerals in foods (Chelating capacity). This molecule is associated with storage proteins, resulting in relatively high concentrations of phytic acid in commercial soy protein products, such as soy meal (Kwanyuen & Burton, 2005). Soybean phytic acid is composed of myo-inositol-1,2,3,4,5,6-hexakis phosphate, which serves as the storage form of phosphorus in the plant. Recent research has found that this molecule has beneficial attributes, such as its ability to reduce the risks of cancer, heart disease, and diabetes due to its antioxidant effects (Fox & Eberl, 2002). It is important to identify or generate low phytate genotypes with nutritional benefits. The diversity in the accumulation of soybean phytic acid may be attributed to various factors, including growing locations, differential soil phosphorus status, pH, temperature, and the presence of phosphate mineralizing organisms in soils (Pandey et al., 2016).

Various techniques, such as cooking, germination, and fermentation, have decreased or eliminated phytates. Hydrothermal treatment does not break down phytates until heated above 130◦C for 45 to 90 min (Schlemmer et al., 2009). Enzymatic degradation is the most effective way to hydrolyze phytates. Phytates are broken down enzymatically by isolated phytases from microorganisms and intrinsic plant phytases that can be activated through certain bioprocesses like germination (Greiner & Konietzny, 2006). Phytic acid is a potent inhibitor of iron-driven hydroxyl radical formation in foods as it can form a unique iron chelate that becomes catalytically inactive. It protects tissues against oxidative reactions, prevents oxidant damage to human/animal organs and foods, and can be added to juice, soybean oil, and meat products to prevent autoxidation and hydrolysis, prevent product discoloration, maintain food safety, and extend shelf-life (Silva & Bracarense, 2016).

In this sense, Ishiguro et al. (2006) analyzed phytate levels in 27 different varieties of soybeans grown in drained paddy and upland fields. They used the Fourier transfer infrared spectroscopy (FT-IR) method for their analysis. Their research found that the phytate content of 12 varieties was higher in drained paddy fields than in upland fields. This suggests that both environmental and hereditary factors can affect the phytate content of soybeans. Similarly, Kim et al. (2006) evaluated the phytate content in five soybean cultivars and found that the phytate content ranged from 18.81 to 22.28 mg/g. The authors also reported antioxidant activity in these phytates using the DPPH method.

## 2.4. Isoflavones

Isoflavones are beneficial non-nutritive substances that have a positive impact on health. They are known to help prevent and treat chronic diseases like heart disease, cancer, diabetes, and hypertension, as well as other clinical pathologies. Isoflavones belong to the flavonoid group and have a flavone nucleus's basic structure, consisting of two benzene rings linked with heterocyclic pyrene (Setchell & Cassidy, 1999). They are mainly found as glycosides in vegetables and are polar and soluble in water. In soybeans, the primary isoflavones are Genistein and Daidzein, which are present in a concentration of 0.1-3.0 mg/g (Alves et al., 2010). These compounds have been found to possess anticancer properties and inhibit the growth of cancer cells by modulating genes that control cell-cycle progression. Genistein, for example, inhibits the activation of the kappa light polypeptide gene enhancer in B-cells (NF-κB). This signaling pathway helps balance cell survival and programmed cell death (apoptosis) (Adjakly, 2013).

De Lemos (2001) observed that high concentrations of genistein could inhibit tumor growth. In an experiment, estrogen-dependent tumors were transplanted onto athymic nude mice whose ovaries had been removed to eliminate endogenous estrogen activity. The mice were then separated into three groups: positive controls were given a 2-mg estradiol pellet subcutaneously, negative controls were given a standard rodent diet, and treatment mice were given an additional daily diet of genistein 750 ppm. Tumor areas were measured weekly for 12 weeks. Tumor growth in the mice fed genistein was similar to that in mice with the estradiol implant. These anticancer effects are higher in Asian populations who consume large amounts of phytoestrogens, compared to Western populations where soy consumption is lower (Hwang & Choi, 2015).

Mice non-obese diabetics (NOD) are models for human type 1 diabetes or insulin-dependent diabetes (IDDM). These mice were given genistein (0.02%, w/w) and daidzein (0.02%, w/w) as part of their diet. After 9 weeks, their blood glucose levels, plasma biomarkers, liver glucose and lipid regulating enzyme activities, and pancreas immunohistochemistry analysis were examined. The results showed that genistein and daidzein supplements increased the insulin/glucagon ratio and C-peptide level while preserving insulin staining in the β-cells of the pancreas in NOD mice. In the liver, these supplements resulted in lower glucose-6-phosphatase (G6Pase) and phosphoenolpyruvate carboxykinase (PEPCK) activities but increased the activities of two lipogenic enzymes, malic enzyme, and glucose-6-phosphate dehydrogenase (G6PD), compared to the control group. These findings suggest that genistein and daidzein are important in regulating glucose homeostasis in type 1 diabetic mice (Choi et al., 2008).

Soybean germ products derived from hypocotyledon present one of the most concentrated sources of isoflavones (> 20 mg/g). After ingestion, intestinal bacteria hydrolyze soybean isoflavones, which may be absorbed or metabolized, providing the beneficial effects already described (Espinosa & Ruperez, 2006).

Wei et al. (1995) evaluated the antioxidant effects of the soybean isoflavone genistein in HL-60 cells and the mouse skin tumorigenesis model. They reported genistein is the most potent inhibitor among TPA-induced $H_2O_2$ formation by dimethyl sulfoxide (DMSO)-differentiated HL-60 cells. On the other hand, dietary administration of 250 ppm genistein for 30 days significantly enhances the activities of antioxidant enzymes in mice's skin and small intestine. Further studies show that genistein significantly inhibits TPA-induced proto-oncogene expression (c-fos) in mouse skin in a dose-dependent manner.

In Japan, Akasa et al. (2002) analyzed daidzein's effect as an inhibitory factor in prostate cancer. The serum level of daidzein was determined and compared in 253 subjects (141 prostate cancer patients and 112 cancer-free controls). The concentrations were lower in the hospitalized subjects.

However, the authors recommended comparing the incidence of daidzein metabolizers among various countries.

## 2.5. Phenolic Compounds

Phenolic compounds naturally comprise several groups, such as simple phenols, phenolic acids, flavonoids, and tannins. These secondary metabolites were related to protecting plants against ultraviolet radiation, pathogens, and herbivores (Harborne & Williams, 2000). Furthermore, phenolic compounds have various biological activities like antioxidative, antimutagenic, antibacterial, atherosclerosis, coronary heart disease, or anticancer effects.

Soybean seed contained various phenolic compounds, and phenolic acids were found to be 28–72% of the total phenol content in soybean seed. The content and profile of phenolic compounds in soybean seed or soy foods were altered by genetic factors (seed size, seed color), environmental factors (temperature, precipitation, storing period), or soy foods processes (heating, steaming, leaching, grinding, and others). According to Chung et al. (2011), the principal phenolic compounds in soybean are mentioned in Table 1.

**Table 1.** Phenolic compounds in soybean

| | Compounds |
|---|---|
| Hydroxybenzoic acid (HBA) derivates (C6-C1) | Gallic acid<br>Protocatechuic acid<br>Gentisic acid<br>p-Hidrixybenzoic acid<br>β-Resorculic acid<br>Vanillic acid<br>Syringic acid<br>Veratric acid<br>Salicylic acid<br>Benzoic acid |
| Hydroxycinnamic acid (HCA) derivates (C6-C3) | Chlorogenic acid<br>o-Coumaric acid<br>m-Coumaric acid<br>p-Coumaric acid<br>Ferulic acid<br>t-Cinnamic acid |

Chung et al., 2011.

These phenolic acids are the remaining phenolic compounds (3000 types) related to color, sensory qualities, and nutritional and antioxidant activity of soybean foods. However, phenolic acids in soybean seeds were related to bitterness/astringency. The daily intakes of phenolic acids and flavonoids were estimated from 20 mg to 1 g in humans, depending on dietary sources and habits (Lee et al., 2008).

According to Dueñas (2012), fermentation can change phenolic compounds. Cho et al. (2009) showed that soybean fermentation by *Bacillus pumilus* produced changes in the esterase activity and phenolic content. An increase in the radical scavenging activity was observed, possibly due to an increased level of isoflavone aglycones and a decrease in its glycosides and phenolic acids. Moreover, phenolic compounds inhibited linolenic acid peroxidation by chelating and reducing $Fe^{3+}$ and neutralizing lipid radicals (Morán et al., 1997). Additionally, these molecules have effectively reduced lipid peroxidation in low-density lipoprotein (LDL) cholesterol molecules, significantly reduced the risk of type II diabetes, and may also possess anti-cancer efficacy (Villegas et al., 2008). Finally, phenolic compounds can act as antioxidants, helping to support the endogenous enzymatic antioxidant systems in cells and scavenge free radicals such as reactive oxygen species (Król-Grzymała & Amarowicz, 2020).

Malenčić et al. (2007) analyzed phenolic compounds' content and antioxidant ability in seeds of twenty soybean hybrids. Phenolic compounds were determined spectrophotometrically after the extraction of seeds with 70% aqueous acetone. The 1,1-diphenyl-2-picrylhydrazyl (DPPH) free radical scavenging activity assay evaluated the antioxidant activity of aqueous acetone extracts. The highest contents of total phenolics were found in Serbian cultivar 1511 and Chinese cultivar LN92-7369, which also displayed the highest total antioxidant activity. Conversely, genotypes poor in phenolics also showed low levels of DPPH-radical scavenging activity. The results suggested that besides protein and oil contents, the phenolic contents should also be considered an important characteristic feature of soybean seeds and a potential selection criterion for antioxidant activity.

Xu and Chang (2008) analyzed the effects of boiling and steaming processes on the phenolic components and antioxidant activities of whole yellow (with yellow seed coat and yellow cotyledon) and black (with black seed coat and green cotyledon) soybeans. As compared to the raw soybeans, all processing methods caused decreases in total phenolic content (TPC), DPPH free radical scavenging activity (DPPH), ferric reducing antioxidant power (FRAP), and oxygen radical absorbing capacity (ORAC) in black soybeans. Pressure steaming caused significant increases in TPC, DPPH, FRAP, and ORAC in yellow soybeans. The pressure steaming treatments caused significant increases in gallic acid and 2,3,4-trihydroxybenzoic acid. In contrast, all treatments caused significant decreases in two predominant phenolic acids (chlorogenic acid and trans-cinnamic acid), and total phenolic acids for yellow and black soybeans. Significant correlations existed between selected phenolic compositions and antioxidant properties of cooked soybeans.

## Conclusion

Soybeans are a globally significant crop due to their high nutritional value, particularly their proteins, lipids, and sugars. However, their secondary metabolites also play a crucial role in promoting human health. These metabolites, synthesized by plants as a defense mechanism, include lectins, saponins, phytates, isoflavones, and phenolic compounds. Together, these molecules exert a synergistic effect to enhance health. Importantly, they remain present in soybean-based foods such as soy milk, soy meat, and other derivatives, and are stable under various conditions, including high temperatures, pH changes, and soaking. Research has shown that the consumption of soybeans and their derivatives is associated with a lower incidence of cancer, particularly among Asian populations. Therefore, further studies are needed to expand the understanding of the health benefits and potential applications of this valuable legume.

## References

Adjakly, M., Ngollo, M., Boiteux, J. P., Bignon, Y. J., Guy, L., & Bernard-Gallon, D. (2013). Genistein and daidzein: different molecular effects on prostate cancer. *Anticancer Research, 33*(1), 39-44. https://pubmed.ncbi.nlm.nih.gov/23267126/.

Akaza, H., Miyanaga, N., Takashima, N., Naito, S., Hirao, Y., Tsukamoto, T. & Mori, M. (2002) Is daidzein non-metabolizer a high risk for prostate cancer? A case-controlled study of serum soybean isoflavone concentration, *Japanese Journal of Clinical Oncology*, 32(8), 296–300, doi:10.1093/jjco/hyf064.

Alves, R. C., Almeida, I. M. C., Casal, S., & Oliveira, M. B. P. P. (2010). Isoflavones in Coffee: Influence of Species, Roast Degree, and Brewing Method. *Journal of agricultural and food chemistry*, *58*(5), 3002-3007. doi:10.1021/jf9039205.

Bljahhina, A., Kuhtinskaja, M., & Krisciunaite, T. (2023). Development of extraction method for determination of saponins in soybean-based yoghurt alternatives: effect of sample pH. *Foods*, *12*(11), 2164. doi:10.3390/foods12112164.

Breeze, M. L., Leyva-Guerrero, E., Yeaman, G. R., Dudin, Y., Akel, R., Brune, P., Claussen, F., Dharmasri, C., Golbach, J., Gou, R., Maxwell, C., Privalle, L., Rogers, H., Liu, K., Shan, G., Yarnall, M., Thiede, D. & Gillikin, N. (2015). Validation of a Method for Quantitation of Soybean Lectin in Commercial Varieties. *Journal of the American Oil Chemists' Society*, 92(8), 1085–1092. doi:10.1007/s11746-015-2679-3.

Cho K. M., Hong S. Y., Math R. K., Lee J. H., Kambiranda D. M., Kim J. M., Islam S. A., Yun M. G., Cho J. J., Lim W. L., Yun H. D. (2009). Biotransformation of phenolics (isoflavones, flavanols and phenolic acids) during the fermentation of cheonggukjang by Bacillus pumilus HY1. *Food Chemistry*, 114, 413–419. doi:10.1016/j.foodchem.2008.09.056.

Choi, M. S., Jung, U. J., Yeo, J. K. M. J., Kim, M. J., & Lee, M. K. (2008). Genistein and daidzein prevent diabetes onset by elevating insulin levels and altering hepatic gluconeogenic and lipogenic enzyme activities in non-obese diabetic (NOD) mice. *Diabetes/Metabolism Research and Reviews*, *24*(1), 74-81. doi:10.1002/dmrr.780.

Crea, F., De Stefano, C., Milea, D., & Sammartano, S. (2008). Formation and stability of phytate complexes in solution. *Coordination Chemistry Reviews*, 252, 1108–1120. doi:10.1016/j. ccr.2007.09.008.

De Lemos, M. L. (2001). Effects of soy phytoestrogens genistein and daidzein on breast cancer growth. *Annals of Pharmacotherapy*, *35*(9), 1118-1121. doi:10.1345/aph.10257.

De Mejia, E. G., Bradford, T., & Hasler, C. (2003). The anticarcinogenic potential of soybean lectin and lunasin. *Nutrition Reviews*, *61*(7), 239-246. doi:10.1301/nr.2003.jul.239-246.

Dong, J., Liang, W., Wang, T., Sui, J., Wang, J., Deng, Z., & Chen, D. (2019). Saponins regulate intestinal inflammation in colon cancer and IBD. *Pharmacological Research*, *144*, 66-72. doi:10.1016/j.phrs.2019.04.010.

Dueñas, M., Hernández, T., Robredo, S., Lamparski, G., Estrella, I., & Muñoz, R. (2012). Bioactive phenolic compounds of soybean (Glycine max cv. Merit): modifications by different microbiological fermentations. *Polish journal of Food and Nutrition sciences*, 62(4). doi: 10.2478/v10222-012-0060-x.

Espinosa-Martos, Irene, and P. Rupérez. Soybean oligosaccharides. Potential as new ingredients in functional food. *Nutricion Hospitalaria* 21.1 (2006): 92-96. https://www.redalyc.org/pdf/3092/309225689016.pdf.

Fox, C. H., & Eberl, M. (2002). Phytic acid (IP6), novel broad spectrum anti-neoplastic agent: a systematic review. *Complementary Therapies in Medicine*, *10*(4), 229-234. doi:10.1016/S0965-2299(02)00092-4.

Graham, T., Graham, M., & Yu, O. (2008). Genomics of secondary metabolism in soybean. In *Genetics and genomics of soybean* (pp. 211-241). New York, NY: Springer New York. doi:10.1007/978-0-387-72299-3_13.

Greiner, R., & Konietzny, U. (2006). Phytase for food application. *Food Technology and Biotechnology*, 44(2), 125–140.

Gou, S. and Chen, C. (2018). Analysis of the bitterness and astringency of soymilk processed from different soybean cultivars. *Journal of Food, Nutrition and Population Health.* 3rd Edition of International Conference on Agriculture & Food Chemistry, Rome Italy. doi: 10.21767/2577-0586-C2-006.

Harborne, J. B., & Williams, C. A. (2000). Advances in flavonoid research since 1992. *Phytochemistry*, 55, 481–504. doi:10.1016/S0031-9422(00)00235-1.

Hassan, S. M., Byrd, J. A., Cartwright, A. L., & Bailey, C. A. (2009). Hemolytic and antimicrobial activities differ among saponin-rich extracts from guar, quillaja, Yucca, and soybean. *Applied Biochemistry and Biotechnology*, 162(4), 1008–1017. doi:10.1007/s12010-009-8838-y.

Hwang, E. S. (2017). Antioxidant and Quinone Reductase Activity of Soyasaponins in Hepa1c1c7 Mouse Hepatocarcinoma Cells. *Preventive Nutrition and Food Science*, *22*(4), 300-305. doi: 10.3746/pnf.2017.22.4.300.

Hwang, K.-A., & Choi, K.-C. (2015). Anticarcinogenic Effects of Dietary Phytoestrogens and Their Chemopreventive Mechanisms. *Nutrition and Cancer*, *67*(5), 796-803. doi:10.1080/01635581.2015.1040516.

Ishiguro, T., Ono T., Wada, T., Tsukamoto, C. & Kono, Y. (2006). Changes in soybean phytate content as a result of field growing conditions and influence on tofu texture, *Bioscience, Biotechnology, and Biochemistry*, 70(4), 874–880, doi:10.1271/bbb.70.874.

Jun, H. S., Kim, S. E., & Sung, M. K. (2002). Protective effect of soybean saponins and major antioxidants against aflatoxin B1-induced mutagenicity and DNA-adduct formation. *Journal of Medicinal Food*, 5(4), 235-240. doi:10.1089/109662002763003393.

Karkle, E. N. L., & Beleia, A. (2010). Effect of soaking and cooking on phytate concentration, minerals, and texture of food-type soybeans. *Food Science and Technology*, *30*, 1056-1060. doi:10.1590/S0101-20612010000400034.

Kim, M. J., Song, Y. J., Kim, H. R., Lee, S. R., Sok, D. E., Kim, S. N., & Kim, M. R. (2009). Polyphenol and phytate contents and their relationship to antioxidative activity in soybeans. *Journal of the East Asian Society of Dietary* Life, 19(6), 975-980.

Kerwin, S. M. (2004). Soy Saponins and the Anticancer Effects of Soybeans and Soy-Based Foods. *Current Medicinal Chemistry - Anti-Cancer Agents*, *4*(3), 263-272. doi:10.2174/1568011043352993.

Król-Grzymała, A., & Amarowicz, R. (2020). Phenolic compounds of soybean seeds from two European countries and their antioxidant properties. *Molecules*, 25(9), 2075. doi:10.3390/molecules25092075.

Kwanyuen, P., & Burton, J. W. (2005). A simple and rapid procedure for phytate determination in soybeans and soy products. *Journal of the American Oil Chemists' Society*, *82*(2), 81-85. doi:10.1007/s11746-005-1046-9.

Lee, S.-J., Kim, J.-J., Moon, H.-I., Ahn, J.-K., Chun, S.-C., Jung, W.-S., et al. (2008). Analysis of isoflavones and phenolic compounds in Korean soybean [Glycine max (L.) Merrill] seeds of different seed weights. *Journal of Agricultural and Food Chemistry*, 56, 2751–2758. doi:10.1021/jf073153f.

Lee, S. Y., Lee, S., Lee, S., Oh, J. Y., Jeon, E. J., Ryu, H. S., & Lee, C. H. (2014). Primary and secondary metabolite profiling of doenjang, a fermented soybean paste during industrial processing. *Food Chemistry*, *165*, 157-166. doi:10.1016/j.foodchem.2014.05.089.

Lin, Peng, Xiujuan Ye, and T. B. Ng. (2008). Purification of melibiose-binding lectins from two cultivars of Chinese black soybeans. *Acta biochimica et Biophysica Sinica* 40 (12): 1029-1038. doi:10.1111/j.1745-7270.2008.00488.x.

Malenčić, D., Popović, M., & Miladinović, J. (2007). Phenolic content and antioxidant properties of soybean (*Glycine max* (L.) Merr.) seeds. *Molecules*, 12(3), 576-581. doi:10.3390/12030576.

Mateos-Aparicio, I., Cuenca, A. R., Villanueva-Suárez, M. J., & Zapata-Revilla, M. A. (2008). Soybean is a promising health source. *Nutricion Hospitalaria*, *23*(4), 305-312. https://pubmed.ncbi.nlm.nih.gov/18604315/.

Moran, J. F., Klucas, R. V., Grayer, R. J., Abian, J., & Becana, M. (1997). Complexes of iron with phenolic compounds from soybean nodules and other legume tissues: prooxidant and antioxidant properties. *Free Radical Biology and Medicine*, 22(5), 861-870. doi:10.1016/S0891-5849(96)00426-1.

Muñoz, R., Arias, Y., Ferreras, J. M., Jiménez, P., Rojo, M. A., & Girbés, T. (2001). Sensitivity of cancer cell lines to the novel non-toxic type 2 ribosome-inactivating protein nigrin b. *Cancer Letters*, *167*(2), 163-169. doi:10.1016/S0304-3835(01)00477-3.

Pan, L., Farouk, M. H., Qin, G., Zhao, Y., & Bao, N. (2018). The influences of soybean agglutinin and functional oligosaccharides on the intestinal tract of monogastric animals. *International Journal of Molecular Sciences*, *19*(2), 554. doi:10.3390/ijms19020554.

Pandey, V., Krishnan, V., Basak, N., Hada, A., Punjabi, M., Jolly, M.& Sachdev, A. (2016). Phytic acid dynamics during seed development and it's composition in yellow and black Indian soybean (*Glycine max* L.) genotypes through a modified extraction and HPLC method. *Journal of Plant Biochemistry and Biotechnology*, *25*, 367-374. doi:10.1007/s13562-015-0348-0.

Schlemmer, U., Frølich, W., Prieto, R. M., & Grases, F. (2009). Phytate in foods and significance for humans: Food sources, intake, processing, bioavailability, protective role and analysis. *Molecular Nutrition and Food Research*, 53, S330–S375. doi:10. 1002/mnfr.200900099.

Setchell, K. D., & Cassidy, A. (1999). Dietary isoflavones: biological effects and relevance to human health. *The Journal of Nutrition*, *129*(3), 758S-767S. doi:10.1093/jn/129.3.758S.

Silva, E. O., & Bracarense, A. P. F. R. L. (2016). Phytic acid: from antinutritional to multiple protection factors of organic systems. *Journal of Food Science*, 81(6), R1357–R1362. doi:10. 1111/1750-3841.13320.

Shi, J., Arunasalam, K., Yeung, D., Kakuda, Y., Mittal, G., & Jiang, Y. (2004). Saponins from edible legumes: chemistry, processing, and health benefits. *Journal of Medicinal Food*, *7*(1), 67-78. doi:10.1089/109662004322984734.

Scheppach, W., Luehrs, H., Melcher, R., Gostner, A., Schauber, J., Kudlich, T. & Menzel, T. (2004). Antiinflammatory and anticarcinogenic effects of dietary fibre. *Clinical Nutrition Supplements*, *1*(2), 51-58. doi:10.1016/j.clnu.2004.09.008.

Tang, S., Li, D., Qiao, S., Piao, X., & Zang, J. (2006). Effects of purified soybean agglutinin on growth and immune function in rats. *Archives of Animal Nutrition*, 60(5), 418–426. doi:10.1080/17450390600884450.

Timilsena, Y. P., Phosanam, A., & Stockmann, R. (2023). Perspectives on saponins: food functionality and applications. *International Journal of Molecular Sciences*, *24*(17), 13538. doi:10.3390/ijms241713538.

Ueno, T., Ohtawa, K., Kimoto, Y., Sakurai, K., Kodera, Y., Hiroto, M. & Inada, Y. (2000). Polyethylene glycol-modified concanavalin A as an effective agent to stimulate anti-tumor cytotoxicity. *Cancer Detection and Prevention*, *24*(1), 100-106. PMID: 10757129.

Van Holle, S. & Van Damme, E. J. (2015). Distribution and evolution of the lectin family in soybean (*Glycine max*). *Molecules*, *20*(2), 2868-2891. doi:10.3390/molecules20022868.

Vasconcelos, I. M., Siebra, E. A., Maia, A. A., Moreira, R. A., Neto, A. F., Campelo, G. J., & Oliveira, J. T. (1997). Composition, toxic and antinutritional factors of newly developed cultivars of Brazilian soybean (Glycine max). *Journal of the Science of Food and Agriculture*, *75*(4), 419-426. doi:10.1002/(SICI)1097-0010(199712)75:4<419::AID-JSFA886>3.0.CO;2-D.

Villegas, R., Gao, Y. T., Yang, G., Li, H. L., Elasy, T. A., Zheng, W., & Shu, X. O. (2008). Legume and soy food intake and the incidence of type 2 diabetes in the Shanghai Women's Health Study. *The American Journal of Clinical Nutrition*, 87(1), 162-167. doi: 10.1093/ajcn/87.1.162.

Wang, R., & Guo, S. (2021). Phytic acid and its interactions: Contributions to protein functionality, food processing, and safety. *Comprehensive Reviews in Food Science and Food Safety*, 20(2), 2081–2105. doi:10.1111/1541-4337.12714.

Wei, H., Bowen, R., Cai, Q., Barnes, S., & Wang, Y. (1995). Antioxidant and antipromotional effects of the soybean isoflavone genistein. *Proceedings of the Society for Experimental Biology and Medicine*, 208(1), 124-130. doi:10.3181/00379727-208-43844.

Xu, B., & Chang, S. K. C. (2008). Total phenolics, phenolic acids, isoflavones, and anthocyanins and antioxidant properties of yellow and black soybeans as affected by thermal processing. *Journal of Agricultural and Food Chemistry*, 56(16), 7165–7175. doi:10.1021/jf8012234.

# Chapter 5

# Bioactive Compounds of *Pisum sativum* L.

**M. A. Porrúa-Ardur**
**and J. C. Ruiz-Ruiz***
Escuela de Nutrición, División de Ciencias de la Salud, Universidad Anáhuac-Mayab, Mérida, Yucatán, México

## Abstract

Legumes are the dry seeds of the Leguminosae family, separated from their pod. They have been cultivated and consumed for centuries as a staple food for humans. Its consumption has been proven to be effective in the prevention of different diseases. Legume seeds, such as soybean, bean, pea, lupin, and lentil, are distributed worldwide due to their low cost, ease of growth, and impact on soil fertility through nitrogen fixation. Additionally, they are rich sources of protein, carbohydrates, water-soluble vitamins and minerals, and other compounds with nutraceutical potential. In this sense, legumes are not only a source of energy and structural compounds, but they can also play a biologically active role as precursors of bioactive compounds with various physiological functions, such as immunomodulatory, antihypertensive, antithrombotic, opioid, etc. There are several alternatives for the bioprocessing of legumes, such as: 1) In vitro enzymatic hydrolysis that uses animal, vegetable, or microbial proteases. 2) Fermentation using specific microorganisms. 3) Germination that activates the metabolism associated with generating a new seedling. Among legumes, *Pisum sativum* L. is the oldest domesticated and second most widely cultivated pulse crop worldwide; it is an annual cool season plant and grows in many parts of the world. Various reports indicate that this legume can be used as a source of compounds with biological activity in developing

* Corresponding Author's Email: jorge.ruizr@anahuac.mx

In: Legumes: Beyond Their Nutritional Value
Editors: C. Jiménez-Martínez, M. F. Juárez-Chairez and M. S. Cid-Gallegos
ISBN: 979-8-89530-440-2

functional foods and nutraceuticals. This chapter deals with the nutritional contribution, bioprocessing, bioactive compounds, and biological activity of *Pisum sativum* L.

**Keywords**: legumes, nutrients, bioprocessing, bioactive compounds, *Pisum sativum* L.

## 1. Introduction

*Pisum sativum*, commonly known as the garden pea, is the third most important food legume in the Fabaceae family, grown worldwide, and is incorporated into human diets as fresh, processed, or dried vegetables and used as a forage crop for animals (FAOSTAT, 2019). It is a legume with high productive value at the agricultural level and an important source of nutrients (Kour et al., 2020; Tulbek et al., 2017). The annual world production of green and dry *P. sativum* amounts to approximately 14.5 million tons and 22 million tons, respectively (FAOSTAT, 2019). The seeds of *P. sativum* provide fundamental nutrients such as soluble and insoluble fiber, proteins rich in tryptophan and lysine, complex carbohydrates, vitamin B, folate, minerals such as calcium, potassium, iron, as well as low content of saturated fat, and sodium (Millar et al., 2019; Kumari and Deka, 2021). Its consumption as part of balanced eating habits decreases the prevalence of colon cancer, the development of type 2 diabetes, coronary conditions, and dyslipidemia (Kour et al., 2020). Both the nutritional quality and the techno-functional properties of the *P. sativum* protein can be improved by various processing methods such as soaking, germination, fermentation, pasteurization, and the incorporation of advanced techniques such as pulsed electric fields or ultrasound-assisted extraction (Kumari and Deka, 2021).

## 2. Protein Content, Techno-Functional Properties and Bioactivity

*P. sativum* contains 14.5-28.5% protein, where the range goes up from 24.4 to 27.4% (Santos et al., 2019). Pea protein is labeled as a non-allergic food substance with remarkably high nutritional value without any modification in the genetic profile (Ge et al., 2020). The major protein of *P. sativum* are globulins and albumins, representing about 80% of total seed protein content.

Albumins represent 18-25% and globulins 55-65% of total proteins. Albumins represent 18-25% and globulins 55-65% of total proteins. All globulins and some albumins are storage proteins used as nitrogen sources for the new embryos after seed germination. Major storage proteins, legumin, vicilin, and convicilin, are globulins and represent 65-85% of total proteins (Choi et al., 2022; Lu et al., 2020). These proteins are classified into 7S (vicilin, convicilin) and 11S fractions (legumin) according to sedimentation properties. Pea proteins are rich in lysine, leucine, and phenylalanine but relatively low in sulfur-containing amino acids, namely methionine and cysteine (Yi et al., 2021).

In the last two decades, enzymatic hydrolysis has been recognized as effective and, from the point of view of safety, as the most appropriate way of modifying proteins of vegetable origin. Peptides generated during partial proteolysis have a smaller molecular size and less compact structure than the original proteins. Peptides contribute to increased techno-functional properties compared to those of unmodified native proteins. Hydrolysis must be carried out under strictly controlled conditions up to a specific degree of hydrolysis to obtain the most suitable techno-functional properties. A limited degree of hydrolysis usually improves solubility and emulsifying and foaming capacities, while excessive hydrolysis decreases some of these functionalities (Barac et al., 2012). The storage proteins of *P. sativum* seeds are very interesting for formulating new food products because of their high nutritive value, non-allergenic character, and good functional properties (Ruiz et al., 2020). In this sense, a good protein product must possess multiple functionalities to perform well in food systems. Protein products' most important techno-functional properties are solubility, emulsification, foaming, and gelation.

As a techno-functional ingredient, *P. sativum* proteins are usually used as flour, concentrates, and isolates. Flour is prepared from shelled and ground seeds, and protein concentrates are produced by air-classification of the *P. sativum* flour (obtained from the milled seeds), which is a dry processing method that blows away the lighter starch granules, thus removing them from the protein (Baro et al., 2015). Concentrates have ≈ 50% content of protein. Protein isolates (> 80%) undergo wet processing in which low molecular weight water-soluble components and salt-soluble proteins are extracted from the flour. Then, the globular proteins are subsequently isolated by a selective precipitation step at the isoelectric point, neutralized, and dried. The final protein content of isolates prepared by isoelectric precipitation is approximately 86%. Protein extraction can occur under alkaline or acidic conditions (Baro et al., 2015).

In general, the techno-functional properties of a protein are affected by numerous factors, which can be classified into two groups: intrinsic and extrinsic factors. The intrinsic factors are amino acid composition and sequence, shape, size, the ratio between hydrophobicity/hydrophilicity, conformation, and reactivity. The extrinsic factors that can affect the techno-functional properties of pure protein include pH, ionic strength, temperature, conformation, the ratio between hydrophobicity/hydrophilicity, and the method of extraction (Baro et al., 2015; Ruiz et al., 2020). Baro et al., 2012 used Papain and commercial protease (*Streptomyces griseus* protease) for protein modification of *P. sativum*. Solubility, emulsifying, and foaming properties were estimated at four different pH values (3.0, 5.0, 7.0, and 8.0). Papain increased solubility, emulsifying properties, and foaming capacity at all pH values. According to this author's proper selection of enzymes, hydrolysis conditions, and genotypes could produce pea protein isolates with desirable functional properties. Techno-functional properties required for a protein product vary due to its specific application in food and food systems (Table 1).

**Table 1.** Some techno-functional properties are performed by functional proteins in food systems

| Techno-functional property | Mechanism of action | Food system |
|---|---|---|
| Solubility | Protein solvation | Beverages |
| Water absorption and binding | Hydrogen bonding of water | Meat, sausages<br>Breads, cakes |
| Emulsification | Formation and stabilization of fat | Emulsification Formation and stabilization of fat |
| Fat absorption | Binding of free fat | Meats, sausages, doughnuts |
| Foaming | Form stable film to entrap gas | Whipped toppings, chiffon |
| Barac et al., 2015. | | |

Various studies suggest that legume proteins, particularly storage proteins like legumins and vicilins, are not merely sources of amino acids; they can induce several physiological effects in animals and humans consuming legume meals or proteins. These physiological effects have been classified into four main categories: (1) adverse immune reactions; (2) modulation of cholesterol and plasma lipid levels; (3) increased intestinal excretion of endogenous proteins; (4) increased excretion of N through the urine and changes in plasma amino acid concentrations (Wang et al., 2022). Adverse immune reactions are relatively rare but are associated with proteins that act as allergens in susceptible individuals, causing responses ranging from mild allergic reactions to life-threatening anaphylaxis (Yu et al., 2016). On the other hand, legume proteins have been shown to modulate cholesterol and plasma lipid levels favorably, potentially reducing the risk of cardiovascular disease (CVD). For instance, soy proteins have been demonstrated to reduce LDL cholesterol levels (Ramdath et al., 2017).

Increased intestinal excretion of endogenous proteins may be linked to legumes' high dietary fiber content. High fiber intake can increase fecal bulk and speed up intestinal transit time, leading to higher excretion of endogenous proteins (Lattimer and Haub, 2010). Meanwhile, increased urinary N excretion and changes in plasma amino acid concentrations have been linked to the high-quality protein content of legumes, which provides all essential amino acids in adequate amounts (Jue et al., 2019). For *P. sativum* protein, recent studies have reported that its bioactivities mainly exhibit antioxidant, antihypertensive, antimicrobial, and antidiabetic characteristics. The antioxidant activities of *P. sativum* proteins, as shown *in vitro*, may confer protection against oxidative stress-related diseases (Nwachukwu and Aluko, 2019).

The gut microbiome, a complex community of microorganisms in the gastrointestinal tract, has emerged as a critical factor involved in the pathophysiology of chronic noncommunicable diseases such as obesity, type-2 diabetes, cancer, and cardiovascular diseases (Iatcu et al., 2021). Current evidence suggests that dietary legume proteins can modulate the gut microbiota, which, in turn, could influence health outcomes, including metabolic health, immune function, and even mental health (Wastyk et al., 2021). While promising, these findings necessitate further research to elucidate the mechanisms underpinning these health effects and to confirm their applicability in humans. Nevertheless, the multifunctionality of legume proteins highlights their potential to contribute to health beyond basic nutrition.

Antioxidant activities have long been recognized in protein hydrolysates owing to their free radical scavenging ability. In this sense, Pownall et al. (2010) made the first report on the antioxidant activity of *P. sativum* protein hydrolysate, using thermolysin, and observed the free radical scavenging ability of peptides with molecular weight <3 kDa. Chelation of prooxidant metal cations presented another possibility to modify the bioactivity of *P. sativum* peptides. Qin et al. (2020) prepared a pea peptide-selenium complex.

They found its improved antioxidant activity compared to the unmodified *P. sativum* peptide, and fragment PPKIYP from the chelated complex was identified. Antihypertensive peptides exhibit an inhibitory mechanism of angiotensin I converting enzyme (ACE), whose downregulation may influence the lowering of blood pressure (Ferreira et al., 2010). One study found that the *P. sativum* protein hydrolyzed inhibited ACE activity both in vitro and when expressed in Caco-2 cells (Bollati et al., 2022).

Antimicrobial biopeptides exhibit mechanisms of action by inhibiting bacteria, fungi, and viruses. A recent study reported that 10 peptide fragments from a *P. sativum* hydrolyzate, which show similarity in charge, isoelectric point, length, and percentage of hydrophobic residues, inhibited the growth of *Escherichia coli* O157:H7 (Mohan et al., 2019). *P. sativum* peptide named psd2 with 47 amino acid residues was found to exhibit antifungal activity against *Aspergillus* and *Fusarium* species, and the authors proved that this activity was based on membrane-targeting mechanism (Amaral et al., 2019). In this sense, some antimicrobial agents directly target a component of bacterial cytoplasmic membranes that can act on Gram-negative and Gram-positive bacteria. Many of these are cyclic peptides with a rigid binding site capable of binding a lipid component. These binding targets antimicrobial agents to bacteria rather than being toxic to host cells (Epand et al., 2016).

Peptides have the potential to influence the pathophysiology of type 2 diabetes through two key mechanisms. First, they can reduce postprandial hyperglycemia—elevated blood glucose levels following carbohydrate consumption—by inhibiting the enzymes α-amylase and α-glucosidase. Second, peptides may act as inhibitors of dipeptidyl peptidase 4 (DPP-4), thereby increasing the levels of glucagon-like peptide-1 (GLP-1) and glucose-dependent insulinotropic polypeptide (GIP). This, in turn, enhances insulin secretion and further reduces postprandial hyperglycemia (Awosika and Aluko, 2019; López-Barón et al., 2018). These mechanisms have been demonstrated in peptides and hydrolysates derived from *P. sativum* (Wei et al., 2019). Additionally, clinical studies have highlighted the role of *P. sativum* proteins and peptides in modulating the gut microbiome, further underscoring their potential in managing type 2 diabetes.

## 3. Carbohydrates/Dietary Fibers, Techno-Functional Properties, and Bioactivity

Legumes are an excellent source of carbohydrates, particularly complex carbohydrates and dietary fibers. These components provide energy, promote digestive health, and have numerous techno-functional properties that make legumes valuable in food processing. Nutritionally, starch greatly impacts the body's glycemic index depending on its digestibility, which is classified as follows: 1) fast-digesting starch, slow-digesting starch, and resistant starch (Singh et al., 2017). Additionally, starch comprises two polymers, one linear (amylose) linear and the other branched (amylopectin). The ratio of these polymers also influences digestibility and the postprandial response to glucose. *P. sativum* starch has an apparent amylose content of 38.2%, which reduces its digestibility and glucose release. This type of starch could be used in food products with a lower glycemic impact (Petropoulou et al., 2020).

Dietary fiber is not one specific compound, but a complex group of many compounds found in plant-based foods, which cannot be fully digested in our gut. Generally, four different groups of dietary fibers are found in foods: 1) Non-starch long carbohydrates. 2) Resistant mid-long carbohydrates such as fructo-oligosaccharides (FOS) and galacto-oligosaccharides (GOS) 3) Resistant starch. 4) Lignin (Stephen et al., 2017). Dietary fiber is also classified as soluble and insoluble fiber. In this sense, lignin, cellulose, and some hemicelluloses fall into insoluble dietary fiber. At the same time, β-glucans, galactomannans, pectin, inulin, and other non-starch polysaccharides are part of soluble dietary fiber (Millar et al., 2019). The physiological effect of dietary fiber is associated with four aspects of its behavior: 1) Modulation of bioavailability by the plant cell walls, 2) Effect of dietary fiber on the rheological and colloidal state of digesta, 3) Binding of dietary fiber with phenolic compounds, bile salts, mineral ions, and digestive enzymes, and 4) Dietary fiber fermentation in the large intestine and the corresponding effect on microbiota composition (Millar et al., 2019).

Both natural antioxidants and dietary fibers are recognized as essential components in promoting human health. In legumes, resistant starch and dietary fiber can function as prebiotics, nourishing beneficial gut bacteria and supporting a healthy gut microbiome. A well-balanced microbiome is linked to numerous health benefits, including improved immune function, better mental health, and reduced risk of chronic diseases (Warman et al., 2022). Natural ingredients that contain both dietary fiber (DF) and antioxidants are referred to as antioxidant dietary fibers. These are defined as dietary fiber concentrates that also contain significant quantities of natural antioxidants, which are intrinsically associated with the dietary fiber matrix, enhancing their potential health benefits. These fibers combine the physiological effects of dietary fiber and antioxidants in a single ingredient, eliminating free radicals and potentially reducing oxidative stress and inflammation in the body (Goni and Saura-Calixto, 2009). *P. sativum* contains 1.2-1.7% pectin, 2.4% cellulose, 1.0% hemicellulose, and 2.5% lignin (Dhingra et al., 2012). The ferric reducing antioxidant power in peapod is 25.9 μmol Trolox equivalents/g and showed a scavenging effect on 2, 2-diphenyl-1-picrylhydrazyl radical as 16.0 mg/mL were also measured (Mateos-Aparicio et al., 2012). Milling, wet fractionation, and dry fractionation processes allow obtaining different types of derivatives from *P. sativum,* such as fractions rich in protein, dietary fiber, and starch, which can be used in the development of functional foods and nutraceuticals (Muneer et al., 2018).

## 4. Phytoconstituents and Bioactivity

Phytoconstituents are non-nutrient active plant compounds that play a crucial role in protecting plants from infections, infestations, and predation. In addition to their protective functions, some phytoconstituents contribute to a plant's organoleptic properties, such as color, aroma, and flavor. These compounds are synthesized through primary and secondary metabolic pathways in plants, and they can be broadly categorized into active drug constituents and inert nondrug constituents (Alamgir, 2018). A wide variety of active components have been identified and classified into 16 or more major groups. The most significant of these include alkaloids, terpenoids, phenols and phenolic glycosides, coumarins and their glycosides, anthraquinones and their glycosides, flavones and flavonoid glycosides (or heterosides), mucilage and gums, tannins, volatile oils, saponins, cardioactive glycosides, and cyanogenic glycosides, among others (Alamgir, 2018).

Oxidative stress is a condition that may occur when there are too many unstable molecules called free radicals in the body and not enough antioxidants to get rid of them. This can lead to cell and tissue damage. Many factors may lead to oxidative stress, including obesity, poor diet, smoking, drinking alcohol, taking certain medicines, and exposure to environmental factors such as radiation, toxins, air pollution, pesticides, and sunlight. Long-term oxidative stress may affect aging and developing chronic inflammation, cancer, and other diseases (Kotha et al., 2022). There are two major mechanisms through which oxidative stress contributes to disease. The first involves the production of reactive species during oxidative stress - particularly •OH, $ONOO^-$ and HOCl - that directly oxidize macromolecules, including membrane lipids, structural proteins, enzymes, and nucleic acids, leading to aberrant cell function and death. The second mechanism of oxidative stress is aberrant redox signaling. Oxidants, particularly $H_2O_2$ generated by cells upon physiological stimulation, can act as second messengers. In oxidative stress, non-physiological production of $H_2O_2$ can cause redox signaling to go awry. Both oxidative stress mechanisms can occur in a single disease (Forman and Zhang, 2021).

Antioxidants can be broadly categorized as 1) Natural and synthetic. 2) Polar and non-polar. 3) Enzymatic and non-enzymatic. 4) Endogenous and exogenous. 5) By the mechanisms in which they are involved. Antioxidants primarily exhibit activities based on three mechanisms: hydrogen atom transfer, single electron transfer, and metal chelation (Granato et al., 2018). They show their activity through three pathways: a) Preventive: prevention of free radical formation and derivatives. b) Interruption: interrupt radical oxidation reactions. c) Inactivation: inactivate free radical/radical derivative reaction products (Mirończuk-Chodakowska et al., 2018).

Numerous phytoconstituents have been identified in *P. sativum*, such as daidzein, genistein, asparaginase, apigenin, and kaempferol, as well as phenolic compounds: catechin, coumaric acid, caffeic acid, vanillic, ferulic, pisatin, protocatechuic acid, proanthocyanidin, steroid phytohormone and tannins (Bello et al., 2018). The concentration of these phytoconstituents in *P. sativum* grains can vary depending on the genotype, cultivation conditions, state of maturity, technological treatment (milling, fractioning, heat treatment, etc.), bioprocesses (hydrolysis, fermentation, germination, etc. (Guo et al., 2019; Zhao et al., 2020).

## Conclusion

*Pisum sativum*, popularly known as the garden pea, has universal cultivation and is recognized as a valuable source of macro and micronutrients. This remarkable legume provides a significant number of proteins, carbohydrates, dietary fibers, vitamins, and minerals, which are essential for human health. However, the benefits of *P. sativum* extend far beyond its nutrient content. It also contains an array of bioactive phytoconstituents that hold promising potential in managing several health concerns. Oxidative stress-related disorders, inflammatory conditions, cancer, blood glucose levels, obesity, and gut health can all be positively influenced by these phytoconstituents. These bioactive compounds, such as phenolic acids, flavonoids, saponins, and fibers, work in many ways, from neutralizing harmful free radicals inhibiting inflammatory pathways to slowing down carbohydrate digestion and feeding the beneficial gut bacteria. In particular, the antioxidative properties of *P. sativum*, attributed to its rich phenolic content, offer a promising strategy for mitigating oxidative stress, a common underlying factor in chronic diseases. Its anti-inflammatory capabilities, credited mainly to flavonoids, could pave the way for innovative strategies for controlling inflammation, a key component in numerous diseases. The presence of anti-carcinogenic phytochemicals like saponins and coumestrol gives a glimpse into the potential of *P. sativum* in cancer prevention strategies. The role of *P. sativum* in glycemic control and weight management can largely be attributed to its high fiber content. Such characteristics suggest that *P. sativum* can contribute significantly to managing metabolic diseases like diabetes and obesity, which are of global concern. Furthermore, the prebiotic effect of the dietary fibers found in *P. sativum* underscores the importance of this legume in promoting gut health. This area garners increasing attention due to the gut's integral role in overall health. The excellent techno-functionality and potential bioactivity of *P. sativum* position this legume as a promising candidate for diverse applications in the food industry. Its suitability extends from conventional food processing, where its texturizing and water-binding properties can be leveraged, to the burgeoning field of functional foods and nutraceuticals. Here, the health-promoting bioactive compounds can be capitalized upon to design foods and supplements to support health and prevent disease. In summary, the widespread cultivation, impressive nutrient profile, versatile techno-functional properties, and potential health-promoting bioactivities collectively underscore the value of *P. sativum*. They spotlight this humble legume as a dietary staple and a valuable resource for health-promoting food products and nutraceuticals. Given this potential, further research is warranted to elucidate the bioactive components' exact mechanisms of action and to confirm these benefits in well-controlled human trials. Such evidence will facilitate the practical application of these findings, guiding the development of innovative, health-promoting food products and nutraceuticals harnessing the power of *P. sativum*.

## References


Alamgir, A. N. M. (2018). Phytoconstituents—Active and inert constituents, metabolic pathways, chemistry and application of phytoconstituents, primary metabolic products, and bioactive compounds of primary metabolic origin. In: Therapeutic use of medicinal plants and their extracts: Vol 2. *Progress in Drug Research*, 74, 25-164. Springer, Cham. doi: 10.1007/978-3-319-92387-1_2.

Amaral, V. S. G., Fernandes, C. M., Felício, M. R., Valle, A. S., Quintana, P. G., Almeida, C. C., Barreto-Bergter, E., Gonçalves, S., Santos, N. C., Kurtenbach, E. (2019). Psd2 pea defensin shows a preference for mimetic membrane rafts enriched with glucosylceramide and ergosterol. *Biochimica et Biophysica Acta (BBA) - Bioenergetics*, 1861,713-728. doi:10.1016/j.bbamem.2018.12.020.

Aranda-Olmedo, I., Ruiz, R., Peinado, M. J., Rubio, L. A. (2017). A pea (*Pisum sativum* L.) seed albumin extract prevents colonic DSS induced dysbiosis in mice. *Journal of Functional Foods*, 35, 279-294. doi: 10.1016/j.jff.2017.05.038.

Awosika, T. O., Aluko, R. E. (2019). Inhibition of the in vitro activities of α-amylase, α-glucosidase, and pancreatic lipase by yellow field pea (*Pisum sativum* L.) protein hydrolysates. *International Journal of Food Science and Technology*, 54, 2021-2034. doi: 10.1111/ijfs.14087.

Barac, M. B., Cabrilo, S., Stanojevic, S., Pesic, M., Pavlicevic, M., Zlatkovic, B., Jankovic, M. (2012). Functional properties of protein hydrolysates from pea (*Pisum sativum* L) seeds. *International Journal of Food Science and Technology*, 47(7), 1457-1467. doi: 10.1111/j.1365-2621.2012.02993.x.

Barac, M. B., Pesic, M. B., Stanojevic, S. P., Kostic, A. Z., Cabrilo, S. B. (2015). Techno functional properties of pea (*Pisum sativum*) protein isolates- A review. *Acta Periodica Technologica*, 46, 1-18. doi: 10.2298/APT1546001B.

Bello, O. A., Ayanda, O. I., Aworunse, O. S., Olukanmi, B. I., Soladoye, M. O., Esan, E. B., Obembe, O. O. (2018). Solaneciobiafrae: An underutilized nutraceutically-important African indigenous vegetable. *Pharmacognosy Reviews*, 12(23), 128-132. doi: 10.4103/phrev.phrev_43_17.

Bollati, C., Xu, R., Boschin, G., Bartolomei, M., Rivardo, F., Li, J., Arnoldi, A., Lammi, C. (2022). Integrated evaluation of the multifunctional DPP-IV and ACE inhibitory effect of soybean and pea protein hydrolysates. *Nutrients*, 14, 2379. doi: 10.3390/nu14122379.

Capuano, E. (2017). The behavior of dietary fiber in the gastrointestinal tract determines its physiological effect, *Critical Reviews in Food Science and Nutrition*, 57:16, 3543-3564. doi: 10.1080/10408398.2016.1180501.

Choi, H., Taghvaei, M., Smith, B., Ganjyal, G. M. (2020). Biochemical analysis of protein compositions among pea (*Pisum sativum*) cultivars grown in the northwest USA. ACS *Food Science and Technology*, 2(7), 1067-1076. doi: 10.1021/acsfoodscitech.1c00460.

Dhingra, D., Michael, M., Rajput, H., & Patil, R. T. (2012). Dietary fibre in foods: A review. *Journal of Food Science and Technology*, 49(3), 255-266. doi: 10.1007/s13197-011-0365-5.

Epand, R. M., Walker, C., Epand, R. F., Magarvey, N. A. (2016). Molecular mechanisms of membrane targeting antibiotics, *Biochimica et Biophysica Acta (BBA) - Biomembranes*, 1858, 5, 980-987. doi: 10.1016/j.bbamem.2015.10.018.

FAOSTAT. (2019). http://www.fao.org/faostat/en/#data/QC.

Ferreira, A. J., Santos, R. A., Bradford, C. N., Mecca, A. P., Sumners, C., Katovich, M. J., Raizada, M. K. (2010). Therapeutic implications of the vasoprotective axis of the renin-angiotensin system in cardiovascular diseases. *Hypertension*, 55, 207-213. doi: 10.1161/HYPERTENSIONAHA.109.140145.

Forman, H. J., Zhang, H. (2021). Targeting oxidative stress in disease: promise and limitations of antioxidant therapy. *Nature Reviews Drug Discovery*, 20, 689-709. doi: 10.1038/s41573-021-00233-1.

Guo, F., Xiong, H., Wang, X., Jiang, L., Yu, N., Hu, Z., Tsao, R. (2019). Phenolics of green pea (*Pisum sativum* L.) hulls their plasma and urinary metabolites, bioavailability, and in vivo antioxidant activities in a rat model. *Journal of Agricultural and Food Chemistry*, 67(43), 11955-11968. doi: 10.1021/acs.jafc.9b04501.

Granato, D., Shahidi, F., Wrolstad, R., Kilmartin, P., Melton, L. D., Hidalgo, F. J., Miyashita, K., Camp, J. V., Alasalvar, C., Ismail, A. B., Elmore, S., Birch, G. G., Charalampopoulos, D., Astley, S. B., Pegg, R., Zhou, P., Finglas, P. (2018). Antioxidant activity, total phenolics, and flavonoids contents: Should we ban *in vitro* screening methods? *Food Chemistry*, 264, 471-475. doi: 10.1016/j.foodchem.2018.04.012.

Iatcu, C. O., Steen, A., Covasa, M. (2021). Gut Microbiota and Complications of Type-2 Diabetes. *Nutrients*. 30;14(1):166. doi: 10.3390/nu14010166.

Jue, L., Klebach, M., Visser, M., Hofman, Z. (2019). Amino acid availability of a dairy and vegetable protein blend compared to single casein, whey, soy, and pea proteins: a double-blind, cross-over trial. Nutrients 11:2613. doi: 10.3390/nu11112613.

Kotha, R. R., Tareq, F. S., Yildiz, E., Luthria, D. L. (2022). Oxidative stress and antioxidants-A critical review on *in vitro* antioxidant assays. *Antioxidants*, 11, 2388. doi: 10.3390/antiox11122388.

Kour, J., Nayik, G. A., ul Haq, R., Anand, N., Alam, M. S., Ramaiyan, B., Banerjee, S. (2020). Pea. In: Nayik, G. A., Gull, A. (eds) Antioxidants in Vegetables and Nuts - Properties and Health Benefits. Springer, Singapore. doi: 10.1007/978-981-15-7470-2_1.

Kumari, T., and Deka, S. C. (2021). Potential health benefits of garden pea seeds and pods: A review. *Legume science*, 3(2), e82. doi: 10.1002/leg3.82.

Lammi, C., Zanoni, C., Aiello, G., Arnoldi, A., Grazioso, G. (2016). Lupin peptides modulate the protein-protein interaction of pcsk9 with the low-density lipoprotein receptor in HepG2 Cells. *Scientific Reports*. 18;6:29931. doi: 10.1038/srep29931.

Lattimer, J. M., Haub, M. D. (2010). Effects of dietary fiber and its components on metabolic health. *Nutrients*. 2(12):1266-89. doi: 10.3390/nu2121266.

López-Barón, N., Sagnelli, D., Blennow, A., Holse, M., Gao, J., Saaby, L., Müllertz, A., Jespersen, B., Vasanthan, T. (2018). Hydrolysed pea proteins mitigate in vitro wheat starch digestibility. *Food Hydrocoll*, 79, 117-126. doi: 10.1016/j.foodhyd. 2017.12.009.

Lu, Z. X., He, J. F., Zhang, Y. C., Bing, D. J. (2020). Composition, physicochemical properties of pea protein and its application in functional foods. *Critical Reviews in Food Science and Nutrition*, 60(15), 2593-2605. doi: 10.1080/10408398.2019.1651248.

Mateos-Aparicio, I., Redondo-Cuenca, A., Villanueva-Suárez, M. J. (2012). Broad bean and pea by-products as sources of fibre-rich ingredients: Potential antioxidant activity measured *in vitro*. *Journal of the Science of Food and Agriculture*, 92(3), 697-703. doi:10.1002/jsfa.4633.

Millar, K. A., Gallagher, E., Burke, R., McCarthy, S., Barry-Ryan, C. (2019). Proximate composition and anti-nutritional factors of fava-bean (*Vicia faba*), green-pea and yellow-pea (*Pisum sativum*) flour. *Journal of Food Composition and Analysis*, 82, 103233. doi:10.1016/j.jfca.2019.103233.

Mirończuk-Chodakowska, I., Witkowska, A. M., Zujko, M. E. (2018). Endogenous non-enzymatic antioxidants in the human body. *Advances in Medical Sciences*, 63, 68-78. doi:10.1016/j.advms.2017.05.005.

Mohan, N. M., Zorgani, A., Jalowicki, G., Kerr, A., Khaldi, N., Martins, M. (2019). Unlocking NuriPep 1653 from common pea protein: a potent antimicrobial peptide to tackle a pan-drug resistant *Acinetobacter baumannii*. *Frontiers in Microbiology*, 10, 2086. doi:10.3389/fmicb.2019.02086.

Muneer, F., Johansson, E., Hedenqvist, M. S., Plivelic, T. S., Markedal, K. E., Petersen, I. L., Sørensen, J. C., Kuktaite, R. (2018). The impact of newly produced protein and dietary fiber rich fractions of yellow pea (*Pisum sativum* L.) on the structure and mechanical properties of pasta-like sheets. *Food Research International*, 106, 607-618, doi:10.1016/j.foodres.2018.01.020.

Nagpal, R., Kumar, M., Yadav, A. K., Hemalatha, R., Yadav, H., Marotta, F., Yamashiro, Y. (2016). Gut microbiota in health and disease: an overview focused on metabolic inflammation. *Benefical Microbes*, 7(2), 181-94. doi:10.3920/bm2015.0062.

Nwachukwu, I. D., Aluko, R. E. (2019). Structural and functional properties of food protein-derived antioxidant peptides. *Journal of Food Biochemistry*. 43(1):e12761. doi:10.1111/jfbc.12761.

Pedrosa, M. M., Varela, A., Domínguez-Timón, F., Tovar, C. A., Moreno, H. M., Borderías, A. J., Díaz, M. T. (2020). Comparison of bioactive compounds content and techno-functional properties of pea and bean flours and their protein isolates. *Plant Foods for Human Nutrition*, 75(4), 642-650. doi:10.1007/s11130-020-00866-4.

Petropoulou, K., Salt, L. J., Edwards, C. H., Warren, F. J., Garcia-Perez, I., Chambers, E. S., Frost, G. S. (2020). A natural mutation in *Pisum sativum* L.(pea) alters starch assembly and improves glucose homeostasis in humans. *Nature Food*, 1(11), 693-704. doi:10.1038/s43016-020-00159-8.

Pownall, T. L., Udenigwe, C. C., Aluko, R. E. (2010). Amino acid composition and antioxidant properties of pea seed (*Pisum sativum* L.) enzymatic protein hydrolysate fractions. *Journal of Agricultural and Food Chemistry*, 58, 4712-4718. doi:10.1021/jf904456r.

Qin, X.-Y., Zhang, J.-T., Li, G.-M., Zhou, M., Gu, R.-Z., Lu, J., Liu, W.-Y. (2020). Structure and composition of a potential antioxidant obtained from the chelation of pea oligopeptide and sodium selenite. *Journal of Functional Foods*, 64, 103619. doi:10.1016/j.jff.2019.103619.

Ramdath, D. D., Padhi, E. M., Sarfaraz, S., Renwick, S., Duncan, A. M. (2017). Beyond the cholesterol-lowering effect of soy protein: a review of the effects of dietary soy and its constituents on risk factors for cardiovascular disease. *Nutrients*. 24;9(4):324. doi:10.3390/nu9040324.

Ruiz, R., Olías, R., Clemente, A., Rubio, L. A. (2020). A pea (*Pisum sativum* L.) seed vicilins hydrolysate exhibits PPARγ ligand activity and modulates adipocyte differentiation in a 3T3-L1 cell culture model. *Foods*. 2020 Jun 16;9(6):793. doi:10.3390/foods9060793.

Santos, C. S., Carbas, B., Castanho, A., Vasconcelos, M. W., Vaz Patto, M. C., Domoney, C., Brites, C. (2019). Variation in pea (*Pisum sativum* L.) seed quality traits defined by physicochemical functional properties. *Foods*, 8(11), 570. doi:10.3390/foods8110570.

Singh, B., Singh, J. P., Shevkani, K., Singh, N., Kaur, A. (2017). Bioactive constituents in pulses and their health benefits. *Journal of Food Science and Technology*, 54(4), 858–870. doi:10.1007/s13197-016-2391-9.

Stephen, A., Champ, M., Cloran, S., Fleith, M., Van Lieshout, L., Mejborn, H., & Burley, V. (2017). Dietary fibre in Europe: Current state of knowledge on definitions, sources, recommendations, intakes, and relationships to health. *Nutrition Research Reviews*, 30(2), 149-190. doi:10.1017/S095442241700004X.

Tulbek, M. C., Lam, R. S. H., Asavajaru, P., Lam, A. (2017). Pea: A sustainable vegetable protein crop. In: Nadathur, S. R., Janitha P. D., Wanasundara, L. S. (eds) Sustainable Protein Sources. Academic Press. doi:10.1016/B978-0-12-802778-3.00009-3.

Warman, D. J., Jia, H., Kato, H. (2022). The potential roles of probiotics, resistant starch, and resistant proteins in ameliorating inflammation during aging (inflammaging). *Nutrients*. 10;14(4):747. doi:10.3390/nu14040747.

Wang, J., Kadyan, S., Ukhanov, V., Cheng, J., Nagpal, R., Cui, L. (2022). Recent advances in the health benefits of pea protein (*Pisum sativum*): bioactive peptides and the interaction with the gut microbiome, *Current Opinion in Food Science*, 48, 100944. doi:10.1016/j.cofs.2022.100944.

Wastyk, H. C., Fragiadakis, G. K., Perelman, D., Dahan, D., Merrill, B. D., Yu, F. B., Topf, M., Gonzalez, C. G., Van Treuren, W., Han, S., Robinson, J. L., Elias, J. E., Sonnenburg, E. D., Gardner, C. D, Sonnenburg, J. L. (2021). Gut-microbiota-targeted diets modulate human immune status. *Cell*. 5;184(16):4137-4153.e14. doi:10.1016/j.cell.2021.06.019.

Wei, Y., Zhang, R., Fang, L., Qin, X., Cai, M., Gu, R., Lu, J., Wang, Y. (2019). Hypoglycemic effects and biochemical mechanisms of Pea oligopeptide on high-fat diet and streptozotocin-induced diabetic mice. *Journal of Food Biochemistry*, 43, e13055. doi:10.1111/jfbc.13055.

Yi, J., Gan, C., Wen, Z., Fan, Y., Wu, X. (2021). Development of pea protein and high methoxyl pectin colloidal particles stabilized high internal phase pickering emulsions for β-carotene protection and delivery. *Food Hydrocolloids*, 113, 106497. doi:10.1016/j.foodhyd.2020.106497.

Yu, W., Freeland, D. M. H., Nadeau, K. C. (2016). Food allergy: immune mechanisms, diagnosis, and immunotherapy. *Nature Reviews Immunology*, 16(12):751-765. doi:10.1038/nri.2016.111.

Zhao, T., Su, W., Qin, Y., Wang, L., & Kang, Y. (2020). Phenotypic diversity of pea (*Pisum sativum* L.) varieties and the polyphenols, flavonoids, and antioxidant activity of their seeds. *Ciencia Rural*, 50(5). doi:10.1590/0103-8478cr20190196.

# Chapter 6

# Phytochemical Compounds Found in *Vicia faba*, *Lens esculenta*, and *Vigna radiata*

**L. Wallander-Compeán**[1]
**M. Rentería-Ortega**[2]
**D. Girón-Orozco**[3]
**and E. B. León-Espinosa**[2,*]
[1]Instituto Politécnico Nacional CIIDIR Unidad Durango, Durango, México
[2]Tecnológico Nacional de México/Tecnológico de Estudios Superiores de San Felipe del Progreso, División de Ingeniería en Industrias Alimentarias, San Felipe del Progreso, México
[3]Facultad de Ciencias Agrícolas de la Universidad Autónoma del Estado de México, Licenciatura en Ingeniero Agrónomo Industrial, Campus Universitario "El Cerrillo," Toluca, México

## Abstract

The Food and Agriculture Organization (FAO) anticipates a doubling of the global population by 2050, which will significantly increase the demand for food. Meeting this demand requires a holistic approach that addresses critical factors such as food security, sustainable development, enhanced agricultural productivity, the eradication of hunger, and improved human health. In this context, leguminous crops, including soybeans, peas, broad beans, and lupines, have garnered attention due to their rising demand, primarily because of their exceptional nutritional profiles. These legumes are rich in proteins, carbohydrates, fatty acids, minerals, and vitamins. Additionally, they contain a variety of bioactive compounds, such as phytates, saponins, tannins, phenolic compounds,

* Corresponding Author's Email: erikab.le@sfelipeprogreso.tecnm.mx

In: Legumes: Beyond Their Nutritional Value
Editors: C. Jiménez-Martínez, M. F. Juárez-Chairez and M. S. Cid-Gallegos
ISBN: 979-8-89530-440-2

lectins, enzyme inhibitors, and isoflavones. The effects of these compounds on human health can be both beneficial and detrimental, depending on their biochemical composition, concentration, and physiological interactions. Furthermore, the biological activity of these compounds can be influenced by factors such as thermal processing, environmental conditions, and the methods used for processing and storing legumes. This chapter seeks to elucidate the bioactive constituents in *Vicia faba*, *Lens esculenta,* and *Vigna radiata*, which have undergone limited scrutiny in scientific investigations.

**Keywords**: polyphenols, legumes, bioactive compounds, nutritional value

## 1. Introduction

There are many reasons why certain people are more likely to develop cardiovascular diseases (CVD), which are the main cause of death worldwide. In developed countries, the rates of CVD are particularly high. Additionally, more people are getting chronic diseases like diabetes, Alzheimer's, cancer, metabolic syndrome, and other age-related conditions. This increases the number of deaths and the amount of money spent on healthcare (Barman et al., 2019). To address these health challenges, people are changing their diets and lifestyles to include more products that promote good health. One example of such products is functional foods (FF), which can provide essential nutrients and improve overall health (Perez-Perez et al., 2018).

Functional foods are defined as any food or food component that provides benefits beyond its nutrients, according to the Institute of Medicine and the National Academy of Sciences (Thomas and Earl, 1994). These foods can naturally contain beneficial components, added during processing, or undergo technological treatments to increase their bioavailability (Davila et al., 2003). Legumes, rich in carbohydrates, proteins, dietary fiber, vitamins, minerals, and bioactive compounds, have been extensively studied for their potential health benefits and are recognized as functional foods (Singh et al., 2017).

Legumes have gained attention for their high content of phytochemicals. These are biologically active compounds produced by plants, including phenolic compounds, flavonoids, phytosterols, alkaloids, lectins, and saponins. These compounds have been shown to possess antioxidant properties, boost the immune system, regulate hormone metabolism, and prevent the formation of blood vessels (Campos-Vega et al., 2010). Additionally, legumes are effective as anticarcinogens, anti-inflammatories, immunomodulators, antimicrobials, and vasodilators (Balasubramaniam et al., 2013).

## 2. Phytochemical Compounds

The term "antinutrient" or "antinutritional factor" refers to compounds that impact the nutritional value of certain foods, particularly seeds, by hindering or inhibiting the assimilation of nutrients of plant origin. These factors can exhibit toxicity or induce undesirable physiological effects, such as flatulence, abdominal distension, and pancreatic affection (Chaparro et al., 2009). Legumes, abundant in bioactive or phytochemical compounds, encompass substances traditionally considered antinutritional factors, which, contemporary research reveals, possess metabolic and physiological effects of interest (Olmedilla-Alonso et al., 2010).

Bioactive compounds in legumes include phytates, galactooligosaccharides, protease inhibitors, lectins, saponins, and phenolic compounds. Antinutritional factors can be categorized as either heat-stable or heat-labile. Thermostable factors encompass antigenic factors, toxic non-protein oligosaccharides, amino acids, saponins, phytoestrogens, cyanogens, and phytates. The most significant are antigenic factors, oligosaccharides, saponins, and phytates. Additionally, thermolabile factors include protease inhibitors (trypsin and chymotrypsin), lectins, goitrogens, and anti-vitamins, with particular emphasis on protease inhibitors and lectins (Chaparro et al., 2009).

This chapter aims to delve into the characteristics of each antinutritional component in *Vicia faba*, *Lens esculenta*, and *Vigna radiata* to elucidate their effects on the human body.

### 2.1. Phytochemical Profiling of Bean (*Vicia faba*)

#### *2.1.1. Phenolic Compounds of Vicia faba*

Legumes are notable reservoirs of bioactive phenolic compounds, which exert significant influence on various physiological and metabolic processes in the human body. These compounds not only serve as bioactive agents but also contribute to the sensory characteristics of foods, playing vital roles in shaping color, taste, and flavor profiles (Singh et al., 2017). Recent research has increasingly explored the diverse physiological effects of polyphenols, including their antihypertensive, antibacterial, and antioxidant properties, as well as their potential to mitigate oxidative stress-related diseases, such as through the inhibition of carcinogenesis (Xu and Chang, 2010). Additionally, phenolic compounds exhibit a wide range of beneficial physiological effects, including anti-allergenic, anti-atherogenic, anti-inflammatory, antimicrobial, antioxidant, anti-thrombotic, cardioprotective, and vasodilatory actions.

In the context of Faba beans (*Vicia faba*), colloquially known as fava horse and broad beans, their global consumption is widespread, spanning regions such as Egypt, Sudan, The Netherlands, Spain, Saudi Arabia, India, China, and Mexico. Faba beans boast a rich nutritional profile featuring proteins, starch, dietary fiber, and fatty acids alongside a diverse repertoire of functional phenolic and flavonoid compounds, including tannins, proanthocyanidins, L-3,4-dihydrophenylalanine (L-dopa), flavonols, flavones, catechins, and phenolic acids, renowned for their intrinsic antioxidant capabilities (Kwon et al., 2018; Kosińska et al., 2011). Their substantial flavan-3-ols content, surpassing 150 mg/kg, is particularly noteworthy (De Pascual et al., 2000). Additionally, fava bean pods contain caffeoyl-L-malic acid, reaching concentrations of up to 100 mg/kg (Dueñas *et al.*, 2006). Phenolic compounds, categorized as microconstituents, have garnered heightened attention due to their health-promoting attributes, predominantly attributed to their antioxidant prowess. These compounds exhibit the potential to reduce the risk of hypertension, cancer, and mutagenic activity, and L-dopa compounds have demonstrated anticancer properties, rendering them plausible candidates for treating Parkinson's disease (Oomah et al., 2011; Kwon et al., 2018).

Extraction methodologies utilizing different solvents have a notable impact on polyphenol content, as demonstrated by Mejri et al. (2018), who found that acetone extraction resulted in the highest phenolic and flavonoid content. The choice of solvent significantly influences antioxidant activity, with acetone and methanol extracts, along with ascorbic acid, showing strong DPPH radical scavenging efficacy. This finding is consistent with previous research, such as Gülcin et al. (2004), which confirmed a positive correlation between total phenolic content and antioxidant activity across various plant species. Bekkara et al. (1998) conducted a comparative analysis of phenolic compound patterns and antioxidant activities in seeds of two broad bean varieties, revealing discernable differences in the distribution of phenolic compounds. Baginsky et al. (2013) explored the phenolic composition of immature seeds across nine *Vicia faba* varieties cultivated in Chile, highlighting considerable diversity in total phenolic and condensed tannins content among the different varieties.

On a global scale, Turco et al. (2016) investigated polyphenol content across *Vicia faba* varieties from Algeria, Australia, Tunisia, and Canada, employing distinct solvent systems. Variability in polyphenol content was evident, underscoring the influence of genetic and environmental factors. Furthermore, Abu-Reidah (2014) identified 104 phenolic compounds in *Vicia faba*, predominantly comprising flavonoids.

In summary, these intricate findings highlight the inherent variability in the phenolic composition of *Vicia faba*, emphasizing its potential as a valuable source of bioactive compounds with complex implications for human health. Table 1 presents some of the key phenolic compounds identified in *Vicia faba*.

**Table 1.** Phenolic compounds identified in the seed extract of *Vicia faba* L.

| Class | Compounds |
|---|---|
| Flavanol monomers | Gallocatechin, (epi) gallocatechin, catechin, epicatechin, di-C-glucoside |
| Prodelphynidins | (epi) gallocatechin-(epi) gallocatechin I, II, III, IV, and V |
| Procyanidin | Procyanidin B, Procyanidin B II, Procyanidin III, Procyanidin B IV |
| Phenolic acid | Salicylic acid O-glucoside, protocatechuic acid hexoside, hydroxyeucomic acid, hydroxybenzyl-malic acid |
| Flavonone | Naringenin 7-glucoside (pruning), pinocembrin |
| Flavone | Myricetine hexose I, quercetin hexose deoxyexose II, kaempferol 3-O-α-L-arabinopyranosyl |
| Flavanol | Myricetin hexose I, quercetin hexose deoxyexose II, myricetin hexose III |
| Dihydrochalcone | Phloretin 3,5-di-C-glucoside |
| Isoflavone | Genisteín, formonetin, daidzein |

## 2.2. Phytochemical Profiling of Lentil (*Lens esculenta*)

The lentil (*Lens esculenta*), an annual plant native to Asia, has undergone diversification from Hindukush to the Mediterranean and the Americas. Renowned for its edible seeds exhibiting diverse shapes and colors, including yellow, red, green, and black (Ganesan and Xu, 2017; Cavalluzzi et al., 2022), the lentil has piqued scientific interest as a promising functional food. Its appeal lies in both its high nutritional value and the presence of bioactive compounds, contributing significantly to disease prevention and the enhancement of human health (Fouad and Rehab, 2015). Consumption of this legume has been associated with mitigating the effects of degenerative diseases, notably diabetes, and cardiovascular disorders (Ganesan and Xu, 2017; Galango et al., 2021; Mustafa et al., 2022).

## 2.3. Nutritional Value of Lentil (*Lens esculenta*)

The nutritional profile of lentils, particularly their seeds, is distinguished by a high energy content (343–356 Kcal), substantial protein content (24.44–25.71%), carbohydrates (60–64.44%), dietary fiber (10.7–31.4%), and simple sugars (2.03–2.86%) (Fouad and Rehab, 2015; Mustafa et al., 2022). These values can vary based on the seed's condition, whether it is raw, germinated, or subjected to a cooking process (Ganesan and Xu, 2017). Furthermore, lentils are a rich source of minerals, including calcium (Ca), iron (Fe), magnesium (Mg), phosphorus (P), potassium (K), sodium (Na), zinc (Zn), and vitamins C, A, and K, niacin, folate. Additionally, they contain monounsaturated fatty acids (0.0179–0.193%) and polyunsaturated fatty acids (0.469–0.526%) (Ganesan and Xu, 2017; Galango et al., 2021).

## 2.4. Bioactive Compounds of Lentil (*Lens esculenta*)

The lentil seed harbors an array of bioactive compounds, encompassing phytosterols, lecithin, polyphenols (both flavonoids and non-flavonoids), isoflavones, triterpenoids, and saponins (Mustafa et al., 2022; Cavalluzzi et al., 2022). Ganesan and Xu (2017) conducted a comprehensive literature review on bioactive compounds within lentils, consolidating the information presented in Table 2. This table delineates various compound groups, functionalities, and quantities per 100 g dry weight.

## 2.5. Phenolic Compounds of Lentil (*Lens esculenta*)

Polyphenols are compounds classified based on the number of phenolic rings and their structural elements or substituents. Two primary groups are distinguished according to the aromatic rings attached to the heterocyclic rings: 1) Flavonoids and 2) Non-flavonoids (Ganesan and Xu, 2017; Irakli et al., 2021).

**Table 2.** Bioactive compounds identified in lentil (*Lens esculenta*)

| Group | Individual component | Weight/100 g of seed | Functionality |
|---|---|---|---|
| Phytosterols | β-sitosterol<br>campesterol<br>stigmasterol | 15.0–24.0 mg<br>15.0 mg<br>20.0 mg | Regulate the membrane fluid |
| Trypsin/protease inhibitors | Bowman–Birk trypsin inhibitors | 3–8 trypsin inhibitor unit (TIU)/mg | Decrease the digestibility of dietary proteins (Anti-nutritional factor); inhibit the cell proliferation in cancer. |
| Lectins | Lectins or hemagglutinins | 12.0 mg | Ability to agglutinate red blood cells RBC and strong stimulators of murine B lymphocyte proliferation. |
| Defensins | Defensins | 8.0 mg | Participate in the development of innate immunity. |
| Dietary Fibers | Fibers<br>Resistant starches | Insoluble fibers (93–99.7 mg/g)<br>Soluble fibers (<7 mg/g)<br>25.4 g (Starches) | Hypocholesterolemic, hypoglycemic, anti-cancer, anti-tumor, antibacterial, and a contributor to gastrointestinal health and gut microbiota. |
| Polyphenols<br>Flavonoids | Flavonols (quercetin and kaempferol). | 0.03-10.85mg<br>0.24-13.20 mg | Antioxidant potential |
| | Flavones, flavanones,<br><br>Proanthocyanidins or condensed tannins (prodelphinidins and procyanidins) | Total phenolic content: 26 mg gallic acid equivalents (GAE/100 g fresh wt); Total flavonoid content: 21 mg catechin equivalents/100 g, and the condensed tannin content of 870 mg catechin equivalents/100 g | Antioxidant activity and potential effect on cardiovascular disease (CVD), diabetes, osteoporosis, and neurodegenerative diseases. |
| | Flavan-3-ols or flavanols (catechin and gallocatechin)<br><br>Anthocyanidins (delphinidin and cyanidin) | 759 mg (GAE)/100 g; glycosides of flavanones: 33.1–186.0 μg; glycosides of flavonols: 9.6–241 μg; dimers procyanidins: 619–1122 μg; trimer procyanidins: 441–498 μg.<br>tetramer procyanidins: 18.5–59.5 μg; galloylated procyanidins: 69.3–123 μg | Antioxidant activity |

**Table 2.** (Continued)

| Group | Individual component | Weight/100 g of seed | Functionality |
|---|---|---|---|
| Polyphenols Non-flavonoids | Hydroxybenzoic acids | 4.5– 28.4 µg | Antioxidant activity and potential effect on diabetes, osteoporosis CVD and neurodegenerative diseases. |
| | Hydroxycinnamic acids (*p*-coumaric acid, ferulic acid and sinapic acid) | Prodelphinidins: 369–725 µg; condensed tannins: 870 mg catechins equivalent | Antioxidant activity |
| | Stilbenoids, trans- resveratrol-3-O-glucoside | Glycosides of trans-resveratrol: 5.5–9.3 µg | Antioxidant activity and potential effect on diabetes, and CVD |
| Phytoestrogens: isoflavones | Formononetin, daidzein, genistein, glycitein, matairesinol, biochanin A, coumestrol, lariciresinol, pinoresinol, secoisolariciresinol, coumestrol | Total isoflavones (9.5 µg), total lignans (26.6 µg) and total phytoestrogens (36.5 µg) | Antioxidant potential |
| Phytate | Phytic acid | 620 mg | Inhibit the proliferation of colorectal cancer. |
| Triterpenoids | Squalene | 0.7 mg | Chemopreventive potential against colorectal cancer. |
| Saponins | Saponins | 25 mg | Hypoglycemic and antidiabetic potential. |

Lentils exhibit a higher polyphenol content compared to other legumes (green peas, chickpeas, cowpeas, yellow peas, mung beans, and peanuts), especially varieties with green, black, or red pigmentation (Fouad and Rehab, 2015; Ahmed et al., 2023). Flavonoids are concentrated in the seed coat, while non-flavonoids are found in the cotyledon. Approximately 35 polyphenols reported in the literature include catechins, procyanidins, flavanols, flavones, flavanones, hydroxybenzoic acids, and hydroxycinnamic acids (Table 2). Concentrations of phenolic compounds or polyphenols may vary among lentil varieties, depending on their color characteristics, environmental conditions, and epigenetic and genotype factors (Fouad and Rehab, 2015; Irakli et al., 2021).

Reported content ranges from 169 to 248 μg/g (DW) and from 1315 to 2381 μg/g DW for derivatives of hydroxybenzoic and hydroxycinnamic acids (in 11 lentil varieties), including sinapic, chlorogenic, caffeic, p-coumaric, m-coumaric, ferulic acids, and syringaldehyde (Xu et al., 2010). Caprioli et al. (2018) reported 16 polyphenols in four different types of lentils, with an average content of 99.2 mg kg−1; they also observed a reduction between 16% and 80% after decortication, cooking, and soaking of the seeds. Recently, five lentil genotypes cultivated in Greece showed ranges of 5.66-6.38 mg GAE/g for total polyphenol concentration; 4.65-5.31 mg CATE/g for total flavonoids; 7.27-8.93 mg PCBE/g for proanthocyanidins, and 2.86-3.2 mg GAE/g for total tannins (Irakli et al., 2021). Similarly, Xia et al. (2023) evaluated four Australian lentil varieties (black, red, green, and brown) and reported the highest concentration of total phenols (0.96 mg GAE/g) in green lentils; red lentils had the highest concentration of total flavonoids (0.06 mg QE/g); black lentils had the highest content of condensed tannins (0.03 mg CE/g), proanthocyanins (0.009 mg CCE/g), and anthocyanins (3.32 mg/100 g); and finally, brown lentils had the highest content of total tannins (2.05 mg TAE/g).

Bioactive compounds such as polyphenols in lentils promote human health, supported by their primarily antidiabetic, cardioprotective, antioxidant, and antimicrobial effects (Cavalluzzi et al., 2022). As mentioned above, this provides valuable information on the importance of including legumes in the human diet, especially when dealing with diseases.

## 2.6. Phytochemical Profiling of Mung Bean (*Vigna radiata*)

The mung bean, a widely consumed legume with particular prominence in Asia, offers numerous health-promoting properties due to its rich composition of proteins, fiber, minerals, vitamins, and various bioactive compounds. These attributes contribute to several health benefits, including the improvement of hyperglycemia, hyperlipidemia, and hypertension, along with the potential to prevent cancer and inhibit melanogenesis. Moreover, mung beans demonstrate hepatoprotective and immunomodulatory effects (Ganesan and Xu, 2018; Hou et al., 2019). In Middle Eastern nations, mung beans are a dietary staple for the entire population, contributing significantly to iron intake, constituting approximately 25% of the total. Acknowledging the presence of antinutritional factors in mung beans, which may impede the optimal utilization of their nutrients and bioactive compounds, is crucial. For instance, phytic acid, forming insoluble complexes with various divalent cations (iron, zinc, calcium, and magnesium), restricts the absorption and utilization of minerals in the small intestine. However, antinutritional effects associated with phytic acid in mung beans can be mitigated through germination, dehulling, and/or cooking (Silva et al., 2013; Ganesan and Xu, 2018). These methods enhance the bioavailability of essential elements, notably iron and zinc, by 3 and 2.4, respectively, compared to raw seeds. This renders mung bean consumption not only viable but also advantageous, harnessing its multifaceted effects, including hypoglycemic, hypolipidemic, anti-hypertensive, anti-cancer, anti-melanogenic, hepatoprotective, immunomodulatory, detoxifying, mental recovery, heatstroke relief, and gastrointestinal disorders regulation properties (Xue et al., 2016; Hou et al., 2019).

## 2.7. Nutritive Value of Mung Bean (*Vigna radiata*)

Mung beans exhibit high nutritional value, which varies depending on whether the seed is raw, germinated, or subjected to a cooking process. Generally, their protein content ranges from 20% to 26% in dry weight, with globulin and albumin being the most abundant storage proteins (Xue et al., 2016). Moreover, seed germination promotes a high amino acid content (18 reported amino acids), the majority of which are essential for the human diet (Silva et al., 2013). Germination also eliminates oligosaccharides that can cause flatulence. Mung beans are rich in starch (62%) but low in calories (327 kcal), making them beneficial for individuals with obesity and diabetes. The mineral and vitamin content are extensive, including calcium, magnesium, iron, phosphorus, potassium, sodium, zinc, copper, manganese, selenium, ascorbic acid, pantothenic acid, thiamine, riboflavin, niacin, vitamins A, E, and K, folate, and vitamin B-6, among others. Additionally, mung beans contain saturated fats (0.348 g/100 g), monounsaturated fats (0.161 g/100 g), and polyunsaturated fats (0.384 g/100 g) (Xue et al., 2016; Ganesan and Xu, 2018).

## 2.8. Bioactive Compounds of Mung Bean (*Vigna radiata*)

Concerning the bioactive compounds present in mung beans, which account for the benefits mentioned above, special attention can be drawn to polyphenols, polysaccharides, and peptides (Silva et al., 2013; Hou et al., 2019). As is the case with all legume species, the concentrations of these bioactive compounds will exhibit variations due to environmental growth conditions, genetic and epigenetic factors, and different varieties, among other factors (Ganesan and Xu, 2017; Hou et al., 2019).

The categories of polyphenols found in mung beans encompass total phenolic acids (ranging from 1.81 to 5.97 mg rutin equivalents/g), total flavonoids (ranging from 1.49 to 1.78 mg catechin equivalents/g), and total tannins (ranging from 1.00 to 5.75 mg/g). Table 3 compiles values reported in the literature for mung bean polyphenols, as Hou et al. (2019) documented.

In a study on the effect of germination under dark and light conditions on the nutritional composition, enzymatic activity, and bioactive compounds of mung beans (*Vigna radiata L.*) variety DX-208, bred and developed by a seed company in southern Vietnam, Van et al. (2020) reported that the duration and germination conditions (light or darkness) significantly affected the content of polyphenols and flavonoids. Regardless of germination conditions, germinated seeds obtained a higher concentration of both bioactive compounds, especially when the germination time was extended to 96 h. At this point, they achieved 7.19 mg FAE/g and 1.92 mg RE/g for total phenols and flavonoids, respectively, compared to non-germinated seeds with only 8.83 mg FAE/g (phenols) and 0.49 mg RE/g (flavonoids). This was attributed to various reasons, including increased biosynthesis and bioaccumulation of phenolic acids during germination as defensive mechanisms to survive environmental stresses. Similarly, Xue et al. (2016) investigated the effect of germination time on a Chinese cultivar of mung beans, reporting a maximum content of total phenols of $\approx$ 6.5 mg GAE/g DM and $\approx$ 5.5 mg RE/g DM for total flavonoids in an optimal germination time of 5 days. They contributed to high antioxidant levels with a maximum of 90.31% elimination against DPPH (2,2-Diphenyl-1-Picrylhydrazyl) and ABTS (2,2'-Azino-bis-(3-ethylbenzothiazoline-6-sulphonic acid) radicals, with the remaining antioxidant power associated with vitamin C, vitamin E, and water-soluble protein.

**Table 3.** Bioactive compounds identified in Mung bean (*Vigna radiata*)

| Class | Subclass | Compound | Content |
|---|---|---|---|
| Flavonoids | Anthocyanins | Cyanidin-3 glucoside | 256.32–476.53 [a] |
| | | Peonidin-3-glucoside | 5.43- 8.42 [a] |
| | | Pelargonidin-3,6 malonylglucoside | 8.53–12.49 [a] |
| | | gonidin-3,6-malonylgluc | 147.62–350.71 [a] |
| | Flavonols | Quercetin | 0.17–16.2[b] |
| | | Myricetin | 0.03–4.26 [b] |
| | | Kaempferol | 0.07–6.13 [b] |
| | Flavanols | Catechin | 4.39–35.36 [b] |
| | Flavones | Vitexin | 17.04–62.37 [b] |
| | | Isovitexin | 22.63–73.64 [b] |
| | | Isovitexin-6”-O—L-glucoside | 170[b] |
| | | Luteolin | 0.36 [b] |
| | Isoflavonoids | Dulcinoside | 13 [b] |
| Phenolic acids | Hydroxycinnamic acid | p-Coumaric acid | 8.17–38.34 [b] |
| | | Caffeic acid | 1.37–38.72 [b] |
| | | t-ferulic acid | 8.02–54.77 [b] |
| | | Chlorogenic acid | 0–26.55 [b] |
| | | Sinapic acid | 22.46 [b] |
| | Hydroxybenzoic acid | Gallic acid | 8.02–54.77 [b] |
| | | Syringic | 0-220.29 [a] |
| | | Gentisic | 25.39-138.45 [a] |

**a**= μg/g, **b**=mg/100 g.

In another study, Bai et al. (2016) extracted stems and leaves (by-products) from mung bean harvests using fractionation extraction with petroleum ether, dichlorobutane, and n-butanol. They reported 22 compounds qualitatively characterized by high-performance liquid chromatography, including 13 flavonoids and 7 lignans already known in the seed. Additionally, 2 isoflavones, vigradiatain (0.0263 g/100 g) and 2'-hydroxy-daidzin (0.0173 g/100 g), were identified, quantified, and reported for the first time. This provides insight into the potential and importance of mung beans as a functional food suitable for consumption in the diet due to the array of benefits provided by its bioactive compounds and nutritional composition.

## 3. Flavonoids

Flavonoids represent a diverse class of plant metabolites characterized by different structural features and oxidation levels within their pyran rings. This group includes aurones, catechins, leucoanthocyanidins (flavan-3,4-diol), flavanones, flavanonols, flavones, flavonols, proanthocyanidins, aurones, chalcones, and isoflavonoids, each distinguished by the position and type of substituents on the pyran ring and benzene ring (Leyva et al., 2011). These compounds play a crucial role in providing color to fruits and flowers, making them relevant to industries involved in pharmaceuticals and food due to their well-documented antioxidant properties (Paniwnyk et al., 2001).

Flavonoids' diverse bioactivities are of particular interest, primarily due to their strong antioxidative properties (*Côté et al.*, 2010). These well-documented antioxidant effects play a crucial role in human nutrition and health. Flavonoids have been associated with potentially slowing the progression of atherosclerosis and cardiovascular diseases by efficiently scavenging reactive oxygen species (ROS). Table 4 outlines a range of biological activities exhibited by various flavonoids extracted from legumes.

**Table 4.** Biological activities of different flavonoids extracted from legumes

| Legumes | Effect and/or mechanisms of action | Reference |
|---|---|---|
| *Phaseolus coccineus* l. | Ayocote beans variety also exhibited the most effective antioxidant activities, which were evaluated by DPPH and ORAC (oxygen radical absorbance capacity) | (Alvarado-López et al., 2019) |
| *Vicia faba* | kaempferol 3-*O*-α-L-rhamnopyranosyl (1→6) (3″-acetyl)-β-D-galactopyranoside and kaempferol 3-*O*-α-L-arabinopyranosyl-5-*O*-α-L-rhamnopyranoside isolated of *Vicia faba* showed lipase and melanogenesis inhibition activities. | (Allam et al., 2018) |
| *Vicia faba* | Showed predominant flavone content such as myricetin, quercetin, apigenin, kaempferol, and luteolin-free aglycones or glucosides. The total content of flavonoids showed ferric-reducing antioxidant activity, the equivalent antioxidant capacity of Trolox, and the radical scavenging activity DPPH. | (Kwon et al., 2018) |

## Conclusion

Legumes contain a variety of bioactive compounds, including saponins, tannins, phenolic compounds, and phytates, which contribute to their biological activities. These compounds have been shown to act as antioxidants, antihypertensives, hypocholesterolemics, immunomodulators, anticancer agents, and anti-obesity agents, among other functions. While these beneficial activities are well-established, it is crucial to investigate the concentration of these compounds and their specific mechanisms of action to better understand how they interact within the body. This chapter explores some of these compounds found in legumes, such as broad beans, lentils, and Phaseolus coccineus, and discusses their concentrations. However, a more comprehensive review is necessary to fully elucidate their effects.

## References

Abu-Reidah, I. M., Contreras, M. M., Arráez-Román, D., Fernández-Gutiérrez, A. (2014). UHPLCESI-QTOF-MS-based metabolic profiling of *Vicia faba* L. (Fabaceae) seeds as a key strategy for characterization in foodomics. Electrophoresis, 35, 1571–1581. doi: 10.1002/elps.201300646.

Ahmed, J., Siddiq, M., & Uebersax, M. A. (2023). *Lentils: Production, processing technologies, products, and nutritional profile*. John Wiley & Sons.

Allam, A. E., Nafady, A. M., Nakagawa, T., Takemoto, N., & Shimizu, K. (2018). Effect of polyphenols from *Vicia Faba* L on lipase activity and melanogenesis. *Natural Product Research*, 32(16), 1920–1925. doi: 10.1080/14786419.2017.1359169.

Alvarado-López, A. N., Gómez-Oliván, L. M., Heredia, J. B., Baeza-Jiménez, R., García-Galindo, H. S., & López-Martínez, L. X. (2019). Nutritional and bioactive characteristics of ayocote bean (*Phaseolus Coccienus* L.): An underutilized legume harvested in Mexico. CYTA - *Journal of Food,* 17(1), 199–206. doi: 10.1080/19476337.2019.1571530.

Baginsky, C., Peña-Neira, Á., Cáceres, A., Hernández, T., Estrella, I., Morales, H., & Pertuzé, R. (2013). Phenolic compound composition in immature seeds of fava bean (*Vicia faba* L.) varieties cultivated in Chile. *Journal of Food Composition and Analysis*, *31*(1), 1-6.

Bai, Y., Xu, Y., Chang, J., Wang, X., Zhao, Y., & Yu, Z. (2016). Bioactives from stems and leaves of mung beans (*Vigna radiata* L.). *Journal of Functional Foods*, 25, 314-322. doi: 10.1016/j.jff.2016.06.009.

Balasubramaniam, V., Mustar, S., Khalid, N. M., Rashed, A. A., Noh, M. F. M., Wilcox, M. D., ... Pearson, J. P. (2013). Inhibitory activities of three Malaysian edible seaweeds on lipase and α-amylase. *Journal of Applied Phycology*, 25(5), 1405–1412. doi: 10.1007/s10811-012-9964-4.

Barman, A., Marak, C. M., Barman, R. M., & Sangma, C. S. (2019). Nutraceutical properties of legume seeds and their impact on human health. in legume seed nutraceutical research. IntechOpen. doi: 10.5772/intechopen.78799.

Bekkara, F., Jay, M., Viricel, M. R., & Rome, S. (1998). Distribution of phenolic compounds within seed and seedlings of two *Vicia faba* cvs differing in their seed tannin content and study of their seed and root phenolic exudations. *Plant and Soil, 203*, 27-36.

Campos-Vega, R., Loarca-Piña, G., & Oomah, B. D. (2010). Minor components of pulses and their potential impact on human health. *Food Research International*. Elsevier. doi: 10.1016/j.foodres.2009.09.004.

Caprioli, G., Nzekoue, F. K., Giusti, F., Vittori, S., & Sagratini, G. (2018). Optimization of an extraction method for the simultaneous quantification of sixteen polyphenols in thirty-one pulse samples by using HPLC-MS/MS dynamic-MRM triple quadrupole. *Food Chemistry, 266*, 490-497. doi: 10.1016/j.foodchem.2018.06.049.

Cavalluzzi, M. M., Lamonaca, A., Rotondo, N. P., Miniero, D. V., Muraglia, M., Gabriele, P., ... De Angelis, E. (2022). Microwave-assisted extraction of bioactive compounds from lentil wastes: antioxidant activity evaluation and metabolomic characterization. *Molecules*, 27, 7471. doi:10.3390/molecules27217471.

Chaparro, D. C., Porrilla, Y., & Elizalde, A. D. (2009). Factores antinutricionales en semillas. *Biotecnología en el Sector Agropecuario y Agroindustrial*, 7. https://revistas.unicauca.edu.co/index.php/biotecnologia/article/view/703.

Côté, J., Caillet, S., Doyon, G., Sylvain, J. F., & Lacroix, M. (2010). Bioactive compounds in cranberries and their biological properties. *Critical Reviews in Food Science and Nutrition*, 50(7), 666–679. doi:10.1080/10408390903044107.

Davila, M., Sangronis, E., & Granito, M. (2003). Leguminosas germinadas o fermentadas: alimentos o ingredientes de alimentos funcionales. *Archivos Latinoamericanos de Nutrición*, 53(4), 348-354.

De Pascual-Teresa, S., Santos-Buelga, C., & Rivas-Gonzalo, J. C. (2000). Quantitative analysis of flavan-3-ols in Spanish foodstuffs and beverages. *Journal of Agricultural and Food Chemistry*, *48*(11), 5331-5337.doi: 10.1021/jf000549h.

Duenas, M., Hernandez, T., & Estrella, I. (2006). Assessment of in vitro antioxidant capacity of the seed coat and the cotyledon of legumes in relation to their phenolic contents. *Food Chemistry*, *98*(1), 95-103. doi:10.1016/j.foodchem.2005.05.052.

Fouad, A. A., & Rehab, F. M. (2015). Effect of germination time on proximate analysis, bioactive compounds and antioxidant activity of lentil (*Lens culinaris* Medik.) sprouts. *Acta scientiarum polonorum. Technologia alimentaria*, *14*(3), 233–246. doi:10.17306/J.AFS.2015.3.25.

Galango, F., Tolve, R., Scarpa, T., Caruso, M. C., Lucini, L., Senizza, B., & Condelli, N. (2021). Extraction kinetics of total polyphenols, flavonoids, and condensed tannins of lentil seed coat: comparison of solvent and extraction methods. *Foods*, 10, 1810. doi:10.3390/foods10081810.

Ganesan, K., & Xu, B. (2017). Polyphenol-rich lentils and their health promoting effects. *International journal of molecular sciences*, *18*(11), 2390. doi:10.3390/ijms18112390.

Ganesan, K., & Xu, B. (2018). A critical review on phytochemical profile and health promoting effects of mung bean (*Vigna radiata*). *Food Science and Human Wellness*, 7(1), 11-33. doi:10.1016/j.fshw.2017.11.002.

Gülçin, İ., Mshvildadze, V., Gepdiremen, A., & Elias, R. (2004). Antioxidant activity of saponins isolated from ivy: α-hederin, hederasaponin-C, hederacolchiside-E and hederacolchiside-F. *Planta medica*, *70*(06), 561-563. doi: 10.1055/s-2004-827158.

Hou, D., Yousaf, L., Xue, Y., Hu, J., Wu, J., Hu, X., Feng, N., & Shen, Q. (2019). Mung Bean (*Vigna radiata* L.): Bioactive Polyphenols, Polysaccharides, Peptides, and Health Benefits. *Nutrients*, 11(6), 1238. doi:10.3390/nu11061238.

Irakli, M., Kargiotidou, A., Tigka, E., Beslemes, D., Fournomiti, M., Pankou, C., & Vlachostergios, D. N. (2021). Genotypic and environmental effect on the concentration of phytochemical contents of lentil (*Lens culinaris* L.). *Agronomy*, 11, 1154. doi: 10.3390/agronomy11061154.

Kosińska, A., Karamać, M., Penkacik, K., Urbalewicz, A., & Amarowicz, R. (2011). Interactions between tannins and proteins isolated from broad bean seeds (*Vicia faba* major) yield soluble and non-soluble complexes. *European Food Research and Technology*, *233*, 213-222. doi:10.1007/s00217-011-1506-9.

Kwon, S. J., Kim, D. G., Kim, J. M., Kang, K. Y., Lee, M. K., Hong, M. J., & Kim, J. B. (2018). Phytochemical compounds and antioxidant activity in the grain of selected faba bean (*Vicia Faba*) genotypes. *Plant Breeding and Biotechnology*, 6(1), 65–73. doi:10.9787/PBB.2018.6.1.65.

Leyva, E., Navarro-Tovar, G., Loredo-Carrillo, S. E., & Santos Díaz, M. del S. (2011). Biosíntesis y actividad biológica de fitoestrógenos y fitoesteroides. *Boletín de la Sociedad Química de México*, 78(210), 35–43. https://bsqm.org.mx/pdf-boletines/V5/N1-3/08.-%20Leyva.pdf.

Mejri, F., Selmi S., Martins, A., Benkhoud, H., Baati, T., Chaabane, H., Njim, L., Serralheiro, M. L. M., Rauter, A. P., Hosni, K. (2018). Broad bean (*Vicia faba* L.) pods: a rich source of bioactive ingredients with antimicrobial, antioxidant, enzyme inhibitory, anti-diabetic, and health-promoting properties. *Food & Function*. 9(4), 2051-2069. doi: 10.1039/c8fo00055g.

Mustafa, A. M., Abouelenein, D., Acquaticci, L., Alessandroni, L., Angeloni, S., Borsetta, G., Caprioli, G., Nzekoue, F. K., Sagratini, G., & Vittori, S. (2022). polyphenols, saponins and phytosterols in lentils and their health benefits: An Overview. *Pharmaceuticals* (Basel, Switzerland), 15(10), 1225. doi:10.3390/ph15101225.

Olmedilla Alonso, B., Farré Rovir, R., Asensio Vegas, C., & Martín Pedrosa, M. (2010). Papel de las leguminosas en la alimentación actual. *Actividad Dietética*. doi:10.1016/S1138-0322(10)70014-6.

Oomah, B. D., Luc, G., Leprelle, C., Drover, J. C. G., Harrison, J. E., Olson, M. (2011). Phenolics, phytic acid, and phytase in Canadiangrown low-tannin faba bean (*Vicia faba* L.) genotypes. *Journal of Agricultural and Food Chemistry*, 59(8), 3763–3771. doi: 10.1021/jf200338b.

Paniwnyk, L., Beaufoy, E., Lorimer, J. P., & Mason, T. J. (2001). The Extraction of rutin from flower buds of Sophora japonica. *Ultrasonics Sonochemistry*, 8(3), 299–301 doi:10.1016/S1350-4177(00)00075-4.

Perez-Perez, L. M., García-Borbón, L., González-Vega, R. I., Rodríguez-Figueroa, J. C., Rosas-Burgos, E. C., Huerta-Ocampo, J. Á., Ruiz-Cruz, S., Wong-Corral, F. J. Balboa-Flores, J., Rueda-Puente, E. O., Del Toro-Sánchez, C. L. (2018). Liberación de compuestos fenólicos ligados en el garbanzo (*Cicer arietinum* L.) utilizando microbiota humana intestinal. *Biotecnia*, 20(3), 146–154. doi:10.18633/biotecnia.v20i3.722.

Silva, L. R., Pereira, M. J., Azevedo, J., Gonçalves, R. F., Valentão, P., Guedes de Pinho, P., & Andrade, P. B. (2013). *Glycine max* (L.) Merr., *Vigna radiata* L. and *Medicago sativa* L. sprouts: A natural source of bioactive compounds. *Food Research International*, 50(1), 167-175. doi:10.1016/j.foodres.2012.10.025.

Singh, B. P., Vij, S., & Hati, S. (2014). Functional Significance of Bioactive Peptides Derived from Soybean. *Peptides*, 54, 171–179. doi:10.1016/j.peptides.2014.01.022.

Singh, B., Singh, J. P., Singh, N., & Kaur, A. (2017). Saponins in pulses and their health-promoting activities: A Review. *Food Chemistry*. doi:10.1016/j.foodchem.2017.04.161.

Thomas, P. R., & Earl, R. (1994). Creating the future of the nutrition and food sciences. *Journal of the American Dietetic Association*, 94(3), 257–259. doi:10.1016/0002-8223(94)90365-4.

Turco, I., Ferretti, G., & Bacchetti, T. (2016). Review of the health benefits of faba bean (*Vicia faba* L.) polyphenols. *Journal of Food & Nutrition Research*, *55*(4).

Van Hung, P., Hoang Yen, N. T., Lan Phi, N. T., Ha Tien, N. P., & Thu Trung, N. T. (2020). Nutritional composition, enzyme activities, and bioactive compounds of mung bean (*Vigna radiata* L.) germinated under dark and light conditions. *LWT*, 133, 110100. doi:10.1016/j.lwt.2020.110100.

Xia, M., Li, M., de Souza, T. S. P., Barrow, C., Dunshea, F. R., & Suleria, H. A. R. (2023). LC-ESI-QTOF-MS2 characterization of phenolic compounds in different lentil (*Lens culinaris* M.) samples and their antioxidant capacity. *Frontiers in bioscience* (Landmark edition), 28(3), 44. doi:10.31083/j.fbl2803044.

Xu, B., & Chang, S. K. C. (2010). Phenolic substance characterization and chemical and cell-based antioxidant activities of 11 lentils grown in the Northern United States. *Journal of Agricultural and Food Chemistry*, 58(3), 1509–1517. https://doi.org/10.1021/jf903532y.

Xue, Z., Wang, C., Zhai, L., Yu, W., Chang, H., Kou, X., & Zhou, F. (2016). Bioactive compounds and antioxidant activity of mung bean (*Vigna radiata* L.), soybean (*Glycine max L.*) and black bean (*Phaseolus vulgaris* L.) during the germination process. *Czech Journal of Food Sciences*, 34, 68-78. doi:10.17221/434/2015-CJFS.

Zeka, K., Ruparelia, K., Arroo, R., Budriesi, R., & Micucci, M. (2017). Flavonoids and their metabolites: prevention in cardiovascular diseases and diabetes. *Diseases*, 5(3), 19. doi:10.3390/diseases5030019.

Zhu, F., Du, B., & Xu, B. (2018). Anti-inflammatory effects of phytochemicals from fruits, vegetables, and food legumes: a review. *Critical Reviews in Food Science and Nutrition*, 58(8), 1260–1270. doi:10.1080/10408398.2016.1251390.

Chapter 7

# Biological Mechanisms of Bioactive Compounds from Legumes

**M. A. Porrúa-Ardura**
**and J. C. Ruiz-Ruiz***
Escuela de Nutrición, División de Ciencias de la Salud, Universidad Anáhuac-Mayab, Mérida, Yucatán, México

## Abstract

The health benefits associated with the consumption of legume seeds have been widely reported across a range of diseases, including cancer, cardiovascular disease, the aging process, immune disorders, diabetes, weight management, osteoporosis, digestive tract diseases, and mental health conditions. Beyond their nutritive compounds, legumes contain numerous phytochemicals, which were traditionally considered anti-nutritional factors but have more recently been recognized as non-nutritional components. While some of these phytochemicals may cause mild discomfort and interfere with protein digestibility and the bioavailability of certain nutrients, such as minerals, the majority are non-toxic to humans. Today, these compounds are acknowledged as bioactive substances capable of exerting health-promoting effects when consumed regularly, thereby reducing the risk of various chronic diseases. Consequently, legumes are now recognized as functional foods. Chronic inflammation and oxidative stress are closely linked to numerous diseases, including diabetes and its complications, hypertension, cardiovascular diseases, neurodegenerative disorders, alcoholic liver disease, chronic kidney disease, cancer, and aging. The mechanism of action of bioactive compounds may be multiple, especially

* Corresponding Author's Email: jorge.ruizr@anahuac.mx

In: Legumes: Beyond Their Nutritional Value
Editors: C. Jiménez-Martínez, M. F. Juárez-Chairez and M. S. Cid-Gallegos
ISBN: 979-8-89530-440-2

in food matrices. Therefore, the interplay between these compounds, the hosts' physiology, and the consequences of their continuous intake should be considered. In analogy with pharmacodynamics, the bioactive compounds should have defined targets and action mechanisms. However, several essential differences arise when considering the heterogeneous nature of the food matrix, the multiplicity of mechanisms, and the variety of responses. This chapter deals with the relationship between proteins and phytoconstituents in legumes and the mechanisms of action through which they exert their physiological effect as antioxidants, anti-inflammatory, and antihypertensive compounds.

**Keywords**: bioactive compounds, diseases, physiology, nutrient-dynamics, mechanisms

## 1. Introduction

Oxidative stress is caused by an imbalance between the production and accumulation of reactive oxygen and nitrogen species in cells and tissues, as well as the detoxification capacity of biological systems (Pizzino et al., 2017). Reactive oxygen and nitrogen species (RONS) perform different functions at the cell signaling level and are generated as metabolism byproducts (Siddeeg et al., 2021). However, the sources of exogenous environmental and xenobiotic stress contribute the most to oxidative stress. In the case of some diseases or stress conditions, the production of antioxidants in the cell is reduced, leading to increases in the RONS/antioxidant ratio, which is identified as oxidative stress (Onyeoziri et al., 2016). The term antioxidant refers to any compound capable of blocking or slowing down the reaction of a substance with RONS, which in turn inhibits the oxidation process. Dietary antioxidants such as peptides, vitamins, flavonoids, and polyphenols can be used from a therapeutic perspective for the treatment of some pathologies linked to oxidative stress, such as cancer, liver disorders, neurological disorders, diabetes mellitus (DM), and cardiovascular disorders (Pizzino et al., 2017; Siddeeg et al., 2021). Legumes are an important source of active dietary antioxidants that may protect the body against oxidative stress. Therefore, there is growing interest in isolating and developing natural antioxidants from plant sources that are both suitable and effective for human use, while minimizing potential side effects (Siddeeg et al., 2021).

## 2. Proteins and Peptides

Generally, bioactive peptides are natural compounds of food or part of protein that are inactive in the precursor molecule. However, they may be active after hydrolysis and can be transported to the active site. Biologically active peptides can also be synthesized chemically and characterized. Peptides have many properties, including antihypertensive, antioxidant, antimicrobial, anticoagulant, and chelating effects (Figure 1).

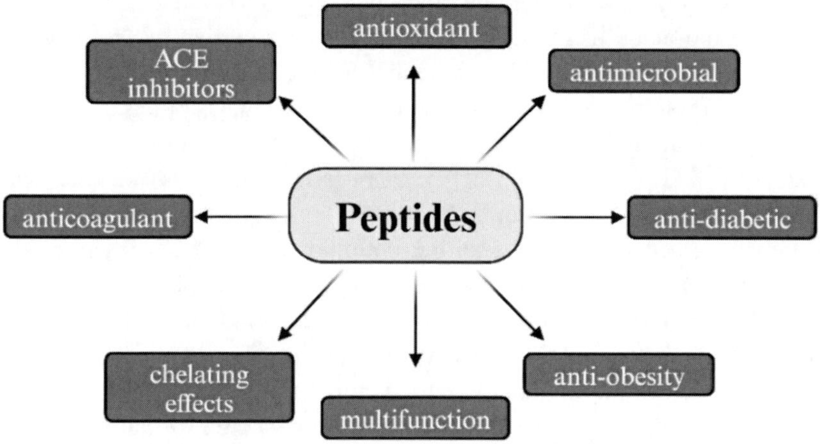


**Figure 1.** Bioactive properties of bioactive peptides (Jakubczyk et al., 2020).

### 2.1. Antioxidant Mechanisms

Molecules can directly and indirectly exhibit antioxidant properties at the cellular level. Direct antioxidant capacity consists of reactive oxygen and nitrogen species uptake by antioxidants through depletion or chemical modifications. On the other hand, indirect antioxidant capacity consists of the ability of antioxidants to increase the activity and expression of endogenous enzymatic and non-enzymatic components (Xing et al., 2021; Zhu et al., 2022). Studies have shown that when the balance between the production of intracellular oxidants and the concentration of antioxidants is altered, exogenous antioxidant peptides can act through direct and indirect pathways, exhibiting intermolecular synergism; for this reason, antioxidant peptides are also known as bifunctional antioxidants (Wang et al., 2018; Liu et al., 2022).

Proteins can inhibit oxidative processes through specific mechanisms, such as enzymes and iron-binding proteins, or less specific mechanisms, such as the effect exerted by phytoconstituents. Both mechanisms contribute to the endogenous antioxidant effect of food and could also enhance their use as antioxidant additives in the food industry. The antioxidant activity of proteins and peptides is due to complex interactions between their ability to inactivate reactive oxygen species, scavenge free radicals, chelate pro-oxidant transition metals, reduce hydroperoxides enzymatically, remove specific oxidants, and alter the physical properties of food systems in a way that inhibits the chemical species associated with oxidation (Elias et al., 2008; Aly and Shahin, 2010).

All amino acids have potential antioxidant properties. A good balance among the various amino acids is essential for the antioxidative properties of plant-based proteins. The most physiologically active peptides have a low molecular weight and short sequences of 2-20 amino acids (Elias et al., 2008; Karami and Akbari-Adergani, 2019). Plant-based foods with a specifically high total protein concentration may contain high levels of amino acids that can transfer electrons, thus exhibiting antioxidant activity. The antioxidant capacity of peptides can be attributed to amino acid composition, sequence, size, and characteristics of side chains of the free amino acids. Plant-derived peptides may exert an antioxidant effect through amino acids acting as metal-chelating agents and hydrogen/electron donors through interactions with free radicals, which may inhibit radical activity or prevent their formation (Aly and Shahin, 2010; Bhandari et al., 2020). Nonetheless, the presence of aromatic (Try, Tyr, and Phe), imidazole (His), and sulfur-containing amino (Cys and Met) groups in peptides results in the development of antioxidative properties. Several amino acids, such as Tyr, Met, His, Lys, Pro, and Trp, reportedly influence antioxidative activity or are accepted as antioxidants and can be predetermined by their amino acid sequence and composition (Bhandari et al., 2020).

Antioxidant mechanisms of proteins and peptides are complex to assess, especially when the *in vitro* and cell culture models cannot be fully correlated. To have results closer to physiological processes, it is possible to use various model organisms to conduct *in vivo* studies on the effects, activity mechanisms, and bioavailability of antioxidant peptides. Mice and rats are the most widely used mammalian models (Aguilar-Toala and Liceaga, 2021). Using some other organisms, such as Drosophila, Caenorhabditis elegans, zebrafish, and turbot is feasible. In recent years, these models have proved ideal for the biological analysis of pharmacologically active compounds due to their short life cycle, ease of culture, and stress response. consistent with that of humans (Li et al., 2020).

In studies carried out with hydrolyzed soybean protein fractions, it was observed that those with high concentrations of His, Tyr, and Phe, exhibit greater radical scavenging activity. Peptides with Trp residues can scavenge free radicals by donating the hydrogen attached to the nitrogen of the indole ring, thus stopping chain reactions. Other studies indicate that the chelating capacity of prooxidant cations of various peptides is related to the presence of His in their sequence (Matemu et al., 2021). In another study carried out with peptides from chickpeas, it was determined that the antioxidant activity of the Asp-Glu-His-Arg-Tyr residue (717.37 Da) was related to the presence of the His residue and its imidazole ring. Additionally, there is evidence that acidic and basic amino acid residues are linked to the ability to chelate metal cations. Studies with soybean peptides indicate that Tyr, Met, His, Lys, and Trp influence antioxidant activity (Nwachukwu and Aluko, 2019).

Proteins, hydrolysates, and peptides derived from legumes have shown in various studies that they could be used as ingredients in the design of functional foods given the potential antioxidant bioactivity that they exhibit through various mechanisms observed, such as free radical scavenging, proton donation, and chelating capacity of metal cations. This antioxidant activity can be applied for preventive management and risk reduction associated with degenerative diseases induced by oxidative stress, inflammatory diseases, hypertension, and cancer.

## 2.2. Anti-Inflammatory Mechanisms

Inflammation is an immune response against injury or disease. This is generally associated with cancer because it involves the interaction of various immune cells that can generate signals for tumor cell proliferation, growth, and invasion. There are two pathways of inflammation-cancer interaction: the extrinsic pathway (inflammation facilitates the development of cancer) and the intrinsic pathway (the genetic changes that cause cancer to stimulate the inflammatory process to promote tumor development) (Crawford, 2014). Anti-inflammatory therapy could reduce or prevent the risk of developing cancer. The peptides lunasin, Val-Pro-Tyr, and Val-Pro-Tyr-glutamyl have anti-inflammatory activity. Lunasin exhibits an anti-inflammatory effect by inhibiting the Akt-mediated NF-kB pathway. Peptides from legumes such as soy regulate various inflammatory markers, such as prostaglandin E2 (PGE2), nitric oxide (NO), induced nitric oxide synthase (iNOS), cyclooxygenase 2 (COX2), cytokines and chemokines (De Mejia and Dia, 2009). Biopeptides from *Phaseolus vulgaris* obtained by enzymatic hydrolysis (alcalase, pepsin, and pancreatin) could also inhibit important inflammatory markers and mediators. These peptides are of low molecular weight and comprise 3 to 11 amino acid residues (Indrati, 2022).

Several bioactive peptides from various sources have been investigated as anti-inflammatory in animal models. This allows the evaluation of the effects of these compounds in more complex physiological conditions. Bean protein hydrolysate with an atherogenic diet showed lower HDL cholesterol (47 mg/dL) than a standard control diet (NC) (68 mg/dL) and atherogenic diet (AD) (55 mg/dL) in male adult BALB/c mice (Gomes et al., 2020). This prevented inflammation and endothelium dysfunction and a decline in oxidative stress, showing an adjuvant effect on lowering atherogenic risk. Soybean protein hydrolysate had a Val-Pro-Tyr peptide sequence which showed an anti-inflammatory effect by reduced MPO activity, decreased gene expression of the pro-inflammatory cytokines TNF-α, IL-6, IL-1β, IFN-γ, and IL-17 in the colon, improved colon histology, and reduced DSS-induced colitis symptoms and weight loss *in vivo* in a mouse model of Dextran Sodium Sulfate (DSS)-induced colitis (Basson et al., 2021). To date, few *in vivo* studies on the anti-inflammatory impact of legume sources have been performed.

## 2.3. Antihypertensive Mechanisms

Peptides with antihypertensive activity have received great attention due to the persistence of hypertension and its associated complications and have even been evaluated in drug interventions. These peptides mainly act as competitive and non-competitive inhibitors of Angiotensin I Converting Enzyme (ACE), a key enzyme of the renin-angiotensin system for regulating blood pressure and electrolyte balance. Peptides with antioxidant and anti-inflammatory properties may also exhibit antihypertensive properties (Majumder and Wu, 2014; Garcés-Rimón et al., 2022). ACE inhibition is the main mechanism by which peptides can modulate renin-angiotensin system function and exert antihypertensive effects. A few *in vivo* studies performed in animals and/or humans have demonstrated that various food-derived peptides could significantly reduce blood pressure through ACE inhibition upon intravenous or oral administration (Majumder and Wu, 2014; Carbonaro and Nucara,

2022). Various methods are used to extract peptides from legumes, e.g., solvent extraction, enzymatic hydrolysis, or fermentation. In a study, thermolysin was applied to obtain peptides from soybeans. Additionally, pepsin and trypsin were used in the second and third steps. An ACE inhibition assay was used in each part, and the $IC_{50}$ value of the hydrolysates was determined to be 53.5 μg/mL, 51.8 μg/mL, and 115.6 μg/mL, respectively. In total, 12 dipeptides, 10 tripeptides, 7 tetrapeptides, 4 pentapeptides, and 1 hexapeptide were identified. Among them, 1 dipeptide (LW) and 5 tripeptides (LLF, LNF, LEF, LSW, IVF) were highlighted, as they were characterized by a predicted $IC_{50}$ value lower than 10 μM using a Quantitative structure-activity relationship model (G u et al., 2013).

Increased production of vasoconstrictor/proinflammatory mediators such as ET-1 and superoxide ($O_2^-$) decreases the bioactivity of the vasodilator NO, leading to endothelial dysfunction. The excessive and/or unregulated generation of reactive oxygen species, such as superoxide and hydrogen peroxide, is one of the main contributors to the pathologies of hypertension (Kibel et al., 2020). In the vascular endothelium, an excess of superoxide causes its reaction with NO to generate peroxynitrite ($ONOO^-$), a highly reactive and toxic molecule that causes the nitration of tyrosine residues of various proteins (Chakrabarti and Wu, 2016; Garcés-Rimón et al., 2022). Antihypertensive peptides derived from food proteins have been shown to improve endothelial functions and generate vasodilation. The three main mechanisms by which food-derived antihypertensive peptides modulate endothelial function are increased production of vasodilatory factors (i.e., NO and prostaglandins), reduced vasoconstrictive factors (i.e., ET -1), and increased antioxidant activity.

## 3. Phenolic Compounds

### 3.1. Antioxidant Mechanisms

Phenolic compounds present in plants can act at molecular and cellular levels as antioxidants. Various epidemiological studies conclude that a diet rich in fruits and vegetables significantly reduces the risk of many diseases related to oxidative stress. Legume seeds are essential to the human diet as they are excellent sources of protein, fiber, minerals, vitamins, and bioactive compounds (Magalhães et al., 2017; Kumar and Goel, 2019). In legumes, the highest concentration of phenolic compounds is found in the pericarp of the seeds. Phenolic compounds act as bioactive phytoconstituents and are related to food's color, flavor, and aroma. They could capture free radicals and interact with proteins. These bioactive phytoconstituents in legume seeds make them suitable for the development of functional foods and nutraceuticals (Singh et al., 2017; Kumar and Goel, 2019).

Phenolic compounds are widely distributed bioactive secondary metabolites found in all higher plants, primarily synthesized through the shikimic acid, pentose phosphate, and phenylpropanoid pathways (Yousefian et al., 2019). Structurally, these compounds contain one or more hydroxyl groups directly attached to an aromatic ring, and they can vary from simple molecules to highly complex polymers. Phenolic compounds are classified into subgroups, including phenolic acids, flavonoids, tannins, and stilbenes, based on the number of attached phenolic hydroxyl groups and the structural elements linking the benzene rings (Zhao et al., 2020).

Antioxidants, as organic or inorganic compounds, have two main mechanisms of action. The primary mechanism is to neutralize free radicals by donating an H atom or transferring an electron. These highly efficient antioxidants are required in very limited amounts to neutralize large numbers of free radicals. The high catalytic properties of these antioxidants demonstrate their ubiquitous presence in nature (Panzella et al., 2020). Phenolic compounds fall into this first category. A neutralization mechanism of pro-oxidant catalysts characterizes secondary antioxidants. Such is the case with pro-oxidant metal ion chelating compounds. These antioxidants generally neutralize a single free radical, so its catalytic activity is quickly exhausted. A third tertiary mechanism is considered, which involves repairing already damaged biomolecules such as DNA or proteins. However, this mechanism is poorly described for dietary antioxidants (Zeb, 2020).

The health benefits of phenolic compounds are attributed to their antioxidant capacity and protective role against chronic diseases induced by free radicals and oxidative stress. Eating foods rich in phenolic compounds decreases the risk of many health problems. Studies have shown that consuming flavonoid-rich foods increases antioxidant activity in blood plasma (Rahman et al., 2021). Other results indicate that the bioavailability of phenolic compounds in beverages and their ability to bind to the lipid fraction of food reduces blood lipid peroxidation. The latter can also be associated with vitamin C in drinks containing phenolic compounds (Bié et al., 2023). The phenolic compounds present in food have potential inhibitory effects on the activity of NADPH oxidase, capture free radicals, and even influence high blood pressure hypertension. Increased intake of phenolic compounds in food may not be the best dietary strategy to overcome oxidative stress, as these compounds can reverse their role and act as pro-oxidants. Therefore, nutritional, toxicological, and medical data must assume a higher dietary intake of phenolic compounds as a safe supplement (Nardini, 2022).

## 3.2. Anti-Inflammatory Mechanisms

Phenolic compounds are a heterogeneous group of compounds derived from the secondary metabolism of plants. Structurally, they comprise at least one aromatic ring to which one or more hydroxyl groups are attached to aromatic or aliphatic structures. Phenolic compounds can be grouped into flavonoids and non-flavonoids. Flavonoids are made up of two aromatic rings linked through an oxygen heterocycle. Depending on the degree of hydrogenation and the substitution of the heterocycle, they can be subclassified into flavonols, flavones, isoflavones, anthocyanins, flavanols, flavanones, etc. Flavonoids usually occur in nature as non-flavonoid glycosides. Benzoic and cinnamic acids are two of the most representative compounds of this type (Ambriz-Pérez et al., 2016).

Plant-derived phytochemicals have been widely used as standardized extracts to treat inflammation and related disorders. Phenolic compounds are essential for suppressing inflammation among them and have been widely referred to in studies due to their anti-inflammatory capacity (Shahidi and Yeo, 2018). In this sense, p-coumaric acid exerts an anti-inflammatory effect by controlling the expression of tumor necrosis factor (TNF-α) in the synovial tissue of arthritic rats. The anti-inflammatory activity of caffeic acid and ellagic acid has also been evaluated; both were supplied to mice mixing them with their diet in a proportion of 2.5 and 5.0%, observing a decrease in the expression of inflammatory mediators such as IL-6, IL-1 -β, tumor necrosis factor (TNF)-α (Pragasam et al., 2016). In other studies, caffeic acid and its derivatives are potent inhibitors of the expression of inducible nitric oxide synthase (iNOS) induced by lipopolysaccharide (LPS) in RAW 264.7 macrophages (Da Cunha et al., 2004). The anti-inflammatory capacity of naringin in cisplatin-induced nephrotoxicity in rats has also been evaluated. The efficacy of hesperidin and rutin in attenuating cisplatin-induced nephrotoxicity in rats has been determined; administration of hesperidin (200 mg/day) or rutin (30 mg/day) for 14 days actively suppressed inflammation in the rats tested. Furthermore, apigenin ameliorated inflammation by reducing p53 activation in human renal proximal tubular epithelial cells. For its part, kaempferol modulated the activities of proinflammatory enzymes, gene expression related to inflammation, and inhibiting transcription factors such as NF-κB (Kamel et al., 2014). Finally, the anti-inflammatory activity of representative flavonoids such as naringenin, quercetin, kaempferol, isorhamnetin, daidzein, and genistein, as well as pelargonidin (anthocyanidin) was investigated. These compounds showed the ability to suppress the iNOS protein by inhibiting nuclear factor-κB (NF-κB), a major transcription factor for iNOS (Hamalainen et al., 2007).

### 3.3. Antihypertensive Mechanisms

The mechanisms underlying the pathophysiology of arterial hypertension are complex, yet a growing body of research has highlighted the involvement of reactive oxygen species (ROS) in the development and maintenance of elevated blood pressure. These studies suggest that oxidative stress plays a significant role in contributing to vascular dysfunction, which is a key factor in the progression of hypertension (Porpino et al., 2016; Nicolás-García et al., 2022). Thus, oxidative stress is involved in the development of hypertension through different mechanisms, which include the inactivation of nitric oxide by superoxide, which leads to endothelial dysfunction and vasoconstriction, and the effect of oxidative stress on the oxidation of low-density lipoproteins. Studies have shown an association between consuming foods and/or beverages rich in phenolic compounds and reducing the risk of cardiovascular diseases (Figueiredo et al., 2017). Their multiple antioxidant properties will mediate the regulation of blood pressure by these phytoconstituents (Nicolás-García et al., 2022).

Arterial hypertension is a chronic non-communicable disease; the therapy used for its management includes antihypertensive drugs. However, blood pressure control remains a problem even with drug therapy. Studies have suggested using phenolic compounds as adjuvant therapy to control arterial hypertension (Gao et al., 2021; De Brito-Alves et al., 2023). The available studies of phenolic compounds related to the control of systemic arterial hypertension include dose, intervention time, and endpoints. Ingestion of a diet containing approximately 425 mg/day of flavonoids distributed in food and associated with the use of the antihypertensive drugs captopril or telmisartan for 6 months reduced blood pressure in hypertensive patients compared with the drugs given alone (De Jesús Romero-Prado et al., 2015; De

Brito-Alves et al., 2023). A study conducted with hypertensive adults and elderly people demonstrated that after two hours of ingestion of 300 ml of cherry juice with high concentrations of anthocyanins (207 mg), there was a decrease in systolic and diastolic blood pressure compared to basal levels (Kent et al., 2016). In another study with hypertensive and elderly adults administered 300 ml of juice of a species of Japanese plum (*Prunus salicina* L.) containing 369 mg of anthocyanins, observed reductions in systolic and diastolic pressures, as well as in mean arterial pressure after consumption. of juice compared to reference values (Igwe et al., 2017). These studies validate the capacity of various phytochemicals, such as flavonoids, also present in legumes.

## Conclusion

Legumes are important beyond their nutritional value and constitute a source of bioactive compounds that exert beneficial biological properties. Among these compounds, proteins and phytoconstituents are notable. In this context, recent works have been compiled on the various mechanisms of action exhibited by these compounds, which can be evaluated by *in vitro*, cellular, and *in vivo* assays. The legumes most reported in these studies are peas, chickpeas, beans, common beans, broad beans, and soybeans. The peptides and the phenolic-type phytoconstituents have shown interesting bioactivities, such as antioxidant, anti-inflammatory, and antihypertensive properties. These biological activities have aroused considerable interest, as they suggest that legumes may have potential therapeutic and preventive applications in chronic noncommunicable diseases, such as cardiovascular disease, diabetes, and certain types of cancer. In addition, these legumes, rich in proteins and bioactive peptides, could improve food's nutritional and functional profile and potentially human health. However, despite the initial promising findings, further research is still needed. Although in vitro studies and cellular assays provide valuable initial insights into the potential mechanisms of action of these bioactive compounds, further *in vivo* studies and human clinical trials are necessary to validate these effects and better understand their bioavailability, metabolism, and potential adverse effects. Additionally, it is crucial to investigate how these bioactive compounds may influence the intestinal microbiota, given its pivotal role in human health. Despite the challenges, the potential of legumes as a rich source of bioactive compounds remains immense. As research progresses, the use of legumes in the development of functional foods and nutraceuticals is likely to expand, offering significant benefits for human health while promoting agricultural and food sustainability.

## References

Aguilar-Toala, J. E., Liceaga, A. M. (2021). Cellular antioxidant effect of bioactive peptides and molecular mechanisms underlying: Beyond chemical properties. *International Journal of Food Science and Technology*, 56:2193-2204. doi: 10.1111/ijfs.14855.

Aly, D. G., Shahin, R. S. (2010). Oxidative stress in lichen planus. *Acta Dermatovenerologica Alpina Pannonica et Adriatica*, 19:3-11.

Ambriz-Pérez, D. L., Leyva-López, N., Gutierrez-Grijalva, E. P., Heredia, J. B. (2016) Phenolic compounds: Natural alternative in inflammation treatment. A Review. *Cogent Food and Agriculture*, 2:1. doi: 10.1080/23311932.2015.1131412.

Basson, A. R., Ahmed, S., Almutairi, R., Seo, B., Cominelli, F. (2021). Regulation of intestinal inflammation by soybean and soy-derived compounds. *Foods*, 4;10(4):774. doi: 10.3390/foods10040774.

Bié, J., Sepodes, B., Fernandes, P. C. B., Ribeiro, M. H. L. (2023). Polyphenols in health and disease: gut microbiota, bioaccessibility, and bioavailability. *Compounds*, 3(1):40-72. doi: 10.3390/compounds3010005.

Bhandari, D., Rafiq, S., Gat, Y., Gat, P., Waghmare, R., Kumar, V. (2020). A review on bioactive peptides: Physiological functions, bioavailability, and safety. *International Journal of Peptide Research and Therapeutics*, 26:139-150. doi: 10.1007/s10989-019-09823-5.

Carbonaro, M., Nucara, A. (2022). Legume proteins and peptides as compounds in nutraceuticals: a structural basis for dietary health effects. *Nutrients,* 11;14(6):1188. doi: 10.3390/nu14061188.

Chakrabarti, S., Wu, J. (2016). Bioactive peptides on endothelial function, *Food Science and Human Wellness*, 5(1):1-7. doi: 10.1016/j.fshw.2015.11.004.

Crawford S. (2014). Anti-inflammatory/antioxidant use in long-term maintenance cancer therapy: A new therapeutic approach to disease progression and recurrence. *Therapeutic Advances in Medical Oncology*, 6:52-68. doi: 10.1177/1758834014521111.

Da Cunha, F. M., Duma, D., Assreuy, J., Buzzi, F. C., Niero, R., Campos, M. M., Calixto, J. B. (2004). Caffeic acid derivatives: *In vitro* and *in vivo* anti-inflammatory properties. *Free Radical Research,* 38:1241-1253. doi: 10.1080/10715760400016139.

De Brito-Alves, J. L., Alves-Brasil, J. M., Araújo-Maia, L., Da Costa-Lima, M., Brito-Sampaio, K., Leite de Souza, E. (2023). Phenolic compounds in hypertension: Targeting gut-brain interactions and endothelial dysfunction, *Journal of Functional Foods*, 104:105531. doi: 10.1016/j.jff.2023.105531.

De Jesús Romero-Prado, M. M., Curiel-Beltrán, J. A., Miramontes-Espino, M. V., Cardona-Muñoz, E. G., Rios-Arellano, A., Balam-Salazar, L. B. (2015). Dietary flavonoids added to pharmacological antihypertensive therapy are effective in improving blood pressure. *Basic and Clinical Pharmacology and Toxicology*, 117(1):57-64. doi: 10.1111/bcpt.12360.

De Mejia, E. G., Dia, V. P. (2009). Lunasin and lunasin-like peptides inhibit inflammation through suppression of NFkB pathway in the macrophage. *Peptides*, 30:2388-2398. doi: 10.1016/j.peptides.2009.08.005.

Elias, R. J., Kellerby, S. S., Decker, E. A. (2008) Antioxidant activity of proteins and peptides. *Critical Reviews in Food Science and Nutrition*, 48:5,430-441. doi: 10.1080/10408390701425615.

Figueiredo, E. A., Alves, N. F. B., Monteiro, M. M. O., Cavalcanti, C. O., Silva, T. M. S. D., Silva, T. M. G. D., Braga, V. A., Oliveira, E. J. (2017). antioxidant and antihypertensive effects of a chemically defined fraction of syrah red wine on spontaneously hypertensive rats. *Nutrients*, 3;9(6):574. doi: 10.3390/nu9060574.

Gao, Q., Xu, L., Cai, J. (2021). New drug targets for hypertension: A literature review. *Biochimica et Biophysica Acta. Molecular basis of disease*, 1867(3):166037. doi: 10.1016/j.bbadis.2020.166037.

Garcés-Rimón, M., Morales, D., Miguel-Castro, M. (2022). Potential role of bioactive proteins and peptides derived from legumes towards metabolic syndrome. *Nutrients* 14(24):5271. doi: 10.3390/nu14245271.

Gomes, M. J. C., Lima, S. L. S., Alves, N. E. G., Assis, A., Moreira, M. E. C., Toledo, R. C. L., Rosa, C. O. B., Teixeira, O. R., Bassinello, P. Z., De Mejía, E. G., Martino, H. S. D. (2020). Common bean protein hydrolysate modulates lipid metabolism and prevents endothelial dysfunction in BALB/c mice fed an atherogenic diet. *Nutrition, Metabolism and Cardiovascular Diseases*, 3;30(1):141-150. doi: 10.1016/ j.numecd.2019.07.020.

Gu, Y., Wu, J. (2013). LC-MS/MS coupled with QSAR modeling in characterising of angiotensin I-converting enzyme inhibitory peptides from soybean proteins. *Food Chem*istry, 141:2682-2690. doi: 10.1016/j.foodchem.2013.04.064.

Hamalainen, M., Nieminen, R., Vuorela, P., Heinonen, M., Moilanen, E. (2007). Anti-inflammatory effects of flavonoids: Genistein, kaempferol, quercetin, and daidzein inhibit STAT-1 and NF-κB activations, whereas flavone, isorhamnetin, naringenin, and pelargonidin inhibit only NF-κB activation along with their inhibitory effect on iNOS expression and NO production in activated macrophages. *Mediators of Inflammation*, 2007:45673. doi: 10.1155/2007/45673.

Igwe, E. O., Charlton, K. E., Roodenrys, S., Kent, K., Fanning, K., Netzel, M. E. (2017). Anthocyanin-rich plum juice reduces ambulatory blood pressure but not acute cognitive function in younger and older adults: a pilot crossover dose-timing study. *Nutrition Researc*h, 47:28-43. doi: 10.1016/j.nutres.2017.08.006.

Indrati, R. (2022). Bioactive peptides from legumes and their bioavailability. *IntechOpen*, doi: 10.5772/intechopen.99979.

Jakubczyk, A., Karaś, M., Rybczyńska-Tkaczyk, K., Zielińska, E., Zieliński, D. (2020). Current Trends of Bioactive Peptides-New Sources and Therapeutic Effect. *Foods*, 29,(7):846. doi: 10.3390/foods9070846.

Jeeva, J. S., Sunitha, J., Ananthalakshmi, R., Rajkumari, S., Ramesh, M., Krishnan, R. (2015). Enzymatic antioxidants and its role in oral diseases. *Journal of Pharmacy and Bioallied Sciences*, S2):S331-3. doi: doi: 10.4103/0975-7406.163438.

Kamel, K. M., El-Raouf, A., Ola, M., Metwally, S. A., El-Latif, A., Hekma, A., Elsayed, M. E. (2014). Hesperidin and rutin, antioxidant citrus flavonoids, attenuate cisplatin-induced nephrotoxicity in rats. *Journal of Biochemistry and Molecular Toxic*ology, 28:312-319. doi: 10.1002/jbt.21567.

Karami, Z., Akbari-Adergani, B. (2019). Bioactive food derived peptides: A review on correlation between structure of bioactive peptides and their functional properties. *Journal of Food Science and Technology*, 56:535-547. doi: 10.1007/s13197-018-3549-4.

Kent, K., Charlton, K. E., Jenner, A., Roodenrys, S. (2016). Acute reduction in blood pressure following consumption of anthocyanin-rich cherry juice may be doseinterval dependant: A pilot cross-over study. *International Journal of Food Sciences and Nutrition*, 67(1):47-52. doi: 10.3109/09637486.2015.1121472.

Kibel, A., Lukinac, A. M., Dambic, V., Juric, I., Selthofer-Relatic, K. (2020). Oxidative stress in ischemic heart disease. *Oxidative Medicine and Cellular Longevity*, 28:6627144. doi: 10.1155/2020/6627144.

Kumar, N., Goel, N. (2019). Phenolic acids: Natural, versatile molecules with promising therapeutic applications. *Biotechnology Reports,* 24:e00370. doi: 10.1016/ j.btre.2019.e00370.

Li, C., Zhang, B., Liu, C., Zhou, H., Wang, X., Mai, K., He, G. (2020). Effects of dietary raw or *Enterococcus faecium* fermented soybean meal on growth, antioxidant status, intestinal microbiota, morphology, and inflammatory responses in turbot (*Scophthalmus maximus* L.). *Fish Shellfish Immunology*, 100, 261-271. doi: 10.1016/j.fsi.2020.02.070.

Liu, H., Liang, J., Xiao, G., Vargas-De-La-Cruz, C., Simal-Gandara, J., Xiao, J., Wang, Q. (2022). Active sites of peptides Asp-Asp-Asp-Tyr and Asp-Tyr-Asp-Asp protect against cellular oxidative stress. *Food Chemistry*, 366:130626. doi: 10.1016/j.foodchem.2021.130626.

Magalhães, S. C. Q., Taveira, M., Cabrita, A. R. J., Fonseca, A. J. M., Valentão, P., Andrade, P. B. (2017). European marketable grain legume seeds: Further insight into phenolic compounds profiles. *Food Chemistry*, 215:177-184. doi: 10.1016/j.foodchem.2016.07.152.

Majumder, K., Wu, J. (2014). Molecular targets of antihypertensive peptides: understanding the mechanisms of action based on the pathophysiology of hypertension. *International Journal of Molecular Sciences*, 24;16(1):256-283. doi: 10.3390/ijms16010256.

Matemu, A., Nakamura, S., Katayama, S. (2021). Health benefits of antioxidative peptides derived from legume proteins with a high amino acid score. *Antioxidants*, 10(2):316. doi: 10.3390/antiox10020316.

Matsui, R., Honda, R., Kanome, M., Hagiwara, A., Matsuda, Y., Togitani, T., Ikemoto, N., Terashima, M. (2018). Designing antioxidant peptides based on the antioxidant properties of the amino acid side-chains. *Food Chemistry*, 245:750-755. doi: 10.1016/j.foodchem.2017.11.119.

Nardini, M. (2022). Phenolic compounds in food: Characterization and health benefits. *Molecules*. 27(3):783. doi: 10.3390/molecules27030783.

Nicolás-García, M., Jiménez-Martínez, C., Perucini-Avendaño, M., Hildeliza Camacho-Díaz, B., Ruperto Jiménez-Aparicio, A., Dávila-Ortiz, G. (2022). Phenolic compounds in legumes: composition, processing, and gut health. *IntechOpen*. doi: 10.5772/intechopen.98202.

Nwachukwu, I. D., Aluko, R. E. (2019). Structural and functional properties of food protein derived antioxidant peptides. *Journal of Food Biochemistry*, 43, e12761. doi: 10.1111/jfbc.12761.

Onyeoziri, U. P., Romanus, E. N., Onyekachukwu, U. I. (2016). Assessment of antioxidant capacities and phenolic contents of nigerian cultivars of onions (*Allium cepa* L) and garlic (*Allium sativum* I). *Pakistan Journal of Pharmaceutical Sciences,* 29(4):1183-1188.

Panzella, L., Moccia, F., Nasti, R., Marzorati, S., Verotta, L., Napolitano, A. (2020). Bioactive phenolic compounds from agri-food wastes: An update on green and sustainable extraction methodologies. *Frontiers in Nutrition*, 7, 60. doi: 10.3389/fnut.2020.00060.

Pizzino, G., Irrera, N., Cucinotta, M., Pallio, G., Mannino, F., Arcoraci, V., Squadrito, F., Altavilla, D., Bitto, A. (2017). Oxidative stress: Harms and benefits for human health. *Oxidative Medicine and Cellular Longevity,* 2017:8416763. doi: 10.1155/2017/8416763.

Porpino, S. K. P., Zollbrecht, C., Peleli, M., Montenegro, M. F., Brandão, M. C. R., Athayde-Filho, P. F., França-Silva, M. S., Larsson, E., Lundberg, J. O., Weitzberg, E. (2016). Nitric oxide generation by the organic nitrate ndbp attenuates oxidative stress and angiotensin ii-mediated hypertension: Organic nitrate and hypertension. *British Journal of Pharmacology,* 173:2290-2302. doi: 10.1111/bph.13511.

Pragasam, S. J., Venkatesan, V., Rasool, M. (2013). Immunomodulatory and anti-inflammatory effect of p-coumaric acid, a common dietary polyphenol, on experimental inflammation in rats. *Inflammation*. 36:169-176. doi: 10.1007/s10753-012-9532-8.

Rahman, M. M., Rahaman, M. S., Islam, M. R., Rahman, F., Mithi, F. M., Alqahtani, T., Almikhlafi, M. A., Alghamdi, S. Q., Alruwaili, A. S., Hossain, M. S., Ahmed, M., Das, R., Emran, T. B., Uddin, M. S. (2021). Role of phenolic compounds in human disease: current knowledge and future prospects. *Molecules*. 30;27(1):233. doi: 10.3390/molecules27010233.

Shahidi, F., Yeo, J. (2018). Bioactivities of phenolics by focusing on suppression of chronic diseases: A review. *International Journal of Molecular Sciences*. 25;19(6):1573. doi: 10.3390/ijms19061573.

Siddeeg, A., AlKehayez, N. M., Abu-Hiamed, H. A., Al-Sanea, E. A., Al-Farga, A. M. (2021). Mode of action and determination of antioxidant activity in the dietary sources: An overview. *Saudi Journal of Biological Sciences,* 28(3):1633-1644. doi: 10.1016/j.sjbs.2020.11.064.

Singh, B., Singh, J. P., Kaur, A., Singh, N. (2017). Phenolic composition and antioxidant potential of grain legume seeds: A review. *Food Research International*, 101:1-16. doi: 10.1016/j.foodres.2017.09.026.

Tawalbeh, D., Al-U'datt, M. H., Wan Ahmad, W. A. N., Ahmad, F., Sarbon, N. M. (2023). Recent advances in *in vitro* and *in vivo* studies of antioxidant, ace-inhibitory and anti-inflammatory peptides from legume protein hydrolysates. *Molecules*, 28(6):2423. doi: 10.3390/molecules28062423.

Xing, L., Fu, L., Hao, Y., Zhang, W. (2021). Dry-cured ham-derived peptide (Asp-Leu-Glu-Glu) exerts cytoprotective capacity in human intestinal epithelial Caco-2 cells. *Antioxidants*, 10, 1354. doi: 10.3390/antiox10091354.

Wang, L., Yu, Q., Han, L., Ma, X., Song, R., Zhao, S., Zhang, W. (2018). Study on the effect of reactive oxygen species-mediated oxidative stress on the activation of mitochondrial apoptosis and the tenderness of yak meat. *Food Chemistry*, 244:394-402. doi: 10.1016/j.foodchem.2017.10.034.

Yousefian, M., Shakour, N., Hosseinzadeh, H., Hayes, A. W., Hadizadeh, F., Karimi, G. (2019). The natural phenolic compounds as modulators of NADPH oxidases in hypertension. *Phytomedicine*, 55, 200-213. doi: 10.1016/j.phymed.2018.08.002.

Zeb, A. (2020). Concept, mechanism, and applications of phenolic antioxidants in foods. *Journal of Food Biochemistry*, 44(9):e13394. doi: 10.1111/jfbc.13394.

Zhao, D., Simon, J. E., Wu, Q. (2020). A critical review on grape polyphenols for neuroprotection: Strategies to enhance bioefficacy. *Critical Reviews in Food Science and Nutrition*, 60(4):597-625. doi: 10.1080/10408398.2018.1546668.

Zhu, Y., Lao, F., Pan, X., Wu, J. (2022). Food Protein-Derived Antioxidant Peptides: Molecular Mechanism, Stability and Bioavailability. *Biomolecules*, 12:1622. doi: 10.3390/biom12111622.

# Chapter 8

# Biological Properties of *Phaseolus vulgaris*: Mechanism of Action, Changes During Processing, and Health Benefits

**M. Nicolás-García[1]**
**J. C. Villalobos-Espinosa[1]**
**J. L. González-Escobar[2]**
**A. Peña-Barrientos[3]**
**and M. Perucini-Avendaño[4,*]**
[1]Tecnológico Nacional de México, ITS de Teziutlán, Ingeniería en Industrias Alimentarias, Teziutlán, Puebla, Mexico
[2]Tecnológico Nacional de México, IT de Ciudad Valles, Carretera Al Ingenio Plan de Ayala Ciudad Valles, San Luis Potosí, SLP, México
[3]Centro de Nanociencias y Micro y Nanotecnologías, Instituto Politécnico Nacional (IPN), Unidad Profesional Adolfo López Mateos, Zacatenco, Mexico City, Mexico
[4]Programa de Investigación y Posgrado en Ciencias de los Alimentos, Facultad de Química, Universidad Autónoma de Querétaro, Santiago de Queretaro, Querétaro, México

## Abstract

The *Phaseolus* genus comprises approximately 70 species and is one of the most popularly consumed legumes worldwide. The common bean (*Phaseolus vulgaris* L.) is a legume of great importance due to its nutritional value and health benefits. Its attributes depend on various factors such as geographic location, genetic and environmental factors,

* Corresponding Author's Email: madeleineperucini@gmail.com

In: Legumes: Beyond Their Nutritional Value
Editors: C. Jiménez-Martínez, M. F. Juárez-Chairez and M. S. Cid-Gallegos
ISBN: 979-8-89530-440-2

and physical characteristics like size, color, and shape (black, yellow, red, brown, brownish, white, purple, and pinto). Beans are an important source of nutrition and contain bioactive compounds such as peptides, phenolic compounds, dietary fiber, unsaturated fatty acids, carotenoids, vitamins, and other components that offer several biological properties. These properties include antioxidant, antidiabetic, antihypertensive, anti-inflammatory, antiproliferative, anticancer, antimicrobial, hypolipidemic, and cardio- and neuroprotective effects. However, the bioactivity of bean components can be positively or negatively affected by the diverse processes to which the bean is subjected, such as post-harvest storage, dehydration, soaking, cooking, germination, fermentation, roasting, and extrusion, among others. Some studies have shown that the antioxidant capacity of certain phytochemicals increases during storage and germination. This chapter describes the beneficial effects of bean compounds on various pathologies, the different biological activities, and mechanisms of action of the bioactive compounds of the bean seed, and how their study has led to the development of functional and nutraceutical foods.

**Keywords**: common bean, bioactive compounds, biological activities, processing

## 1. Introduction

Beans are one of the most economically important legumes in the world as a source of energy and represent 50% of the total dietary legume intake; they provide up to 15% of the daily calories and 36% of the daily complete protein requirements (Qin et al., 2019; Stefanucci et al., 2022; Perera et al., 2023). Their relevance resides in their nutritional composition, which includes proteins, carbohydrates, fiber, minerals, vitamins, and unsaturated fatty acids. In addition, it contains *non-nutritional compounds* such as phytic acid, saponins, and phytochemicals (Finetti et al., 2020; Nicolás-García et al., 2021a; Aravindakshan et al., 2022). The genus Phaseolus comprises species such as *P. vulgaris, P. lunatus, P. coccineus, P. acutifolius,* and *P. dumosus*. More than 40,000 varieties of this genus have been recognized and have different characteristics and nutritional properties. However, *P. vulgaris* usually known as the common bean, is the most commercially important and has about 70 varieties. Their qualities depend on their color, such as black, yellow, red, brown, white, purple, and pinto. Also, the characteristics of growth, maturation, geographical location, and adaptation allow for food biodiversity (Capistrán-Carabarin et al., 2019; Finetti et al., 2020; Alcázar-Valle et al., 2021; Nicolás-García et al., 2021b).

The common bean is considered one of the essential foods for the biological properties exhibited by bioactive compounds like its antidiabetic activity due to counteracting fat accumulation, insulin resistance, and dyslipidemia. In addition, prevention of cardiovascular diseases, antioxidant activity, hypolipidemic, anti-inflammatory, antiproliferative, and prebiotic properties, among others, have been reported (Finetti et al., 2020; Stefanucci et al., 2022). However, the consumption of beans has been changing over time. New processing techniques, such as storage, soaking, dehydration, roasting, cooking, germination, extrusion, and fermentation, have been implemented in addition to the traditional ones. However, bean components can undergo positive or negative changes that influence the product's sensory, nutritional, and functional quality, depending on the processing conditions (Mecha et al., 2019; Cárdenas-Castro et al., 2020).

## 2. Biological Property of Bioactive Compounds in Beans

The presence of bioactive compounds in P. vulgaris, such as bioactive peptides, α-amylase inhibitors, phytohemagglutinin, vitamins, saponins, anthocyanins, flavonoids, polyunsaturated fatty acids, phytosterols, pigments, tocopherols, starch, resistant starch, minerals, and dietary fiber promotes a positive effect on health. Beans participate in decreasing lipid absorption, LDL cholesterol, and chronic degenerative diseases, such as cardiovascular, diabetes, hypertension, osteoporosis, and cancer. Biological properties reported for the common bean include antioxidant, antiproliferative, hypocholesterolemia, anti-inflammatory, and activity on prostate, liver, breast, and colon cancer cells, as shown in Table 1 (Ashraf et al., 2020; Finetti et al., 2020; Martínez-Ávila et al., 2022; Stefanucci et al., 2022).

### 2.1. Antioxidant Activity

The generation of reactive oxygen species (ROS) is a factor in cell transformation and activates pro-oncogenic signaling pathways that induce genetic instability, DNA damage, and mutations. ROS modifies the function of biomolecules such as lipids and proteins, causing oxidations that participate in cell death. The oxidative stress experienced by cells due to the imbalance between the formation of ROS and free radicals has been studied. It has been shown that consuming antioxidants reduces the risk of cell damage and helps prevent cancer. Several authors have evidenced the antioxidant activity of bioactive compounds present in the common bean, such as peptides and phenolic compounds, including flavonoids (Al-Ruwaih et al., 2019; Finetti et al., 2020; Borges-Martínez et al., 2021; Sánchez-Chino et al., 2022).

Several factors influence the antioxidant capacity present in beans, including their variety and color, mainly due to the presence of phytochemicals; types with higher pigmentation contain more phytochemicals than light or white ones (Stefanucci et al., 2022). Diverse phenolic compounds present in beans are involved in the prevention of various chronic degenerative diseases; these include isoflavonoids (genistein) involved in the inhibition of cancer cell growth (breast and prostate), flavonoids such as kaempferol and quercetin that prevent cardiovascular diseases, and anthocyanins (cyanidin 3-glucoside) and phenolic acids (ferulic acid) that have been shown to have potential antioxidant activity (Alcazar-Valle et al., 2020; Stefanucci et al., 2022). Alcazar-Valle et al. (2021) evaluated the antioxidant capacity of three species of *Phaseolus* (*P. coccineus, P. lunatus, and P. vulgaris*), observing that *P. vulgaris* of different colors showed a high content of tannins and anthocyanins and higher antioxidant capacity compared to the other species, assessed by DPPH (89.28%) and ABTS (86.43%) methods compared to 56.47 and 80.57% reported for *P. coccineus*, respectively.

However, other components such as dietary fiber, proteins, and secondary metabolites also benefit the organism (Al-Ruwaih et al., 2019; Ashraf et al., 2020; Sánchez-Chino et al., 2022). Karaman et al. (2021) evaluated bean protein concentrates' bioactive and techno-functional properties, observing a protein content in the flour and concentrate of 22-26.96 and 72.97-77.99%, respectively. An increase in antiradical activity was observed with the DPPH method for samples of flour (14.85-34.31%) and concentrate (23.28-37.32%), depending on the genotype evaluated. The same behavior was observed with the ABTS+ radical scavenging assay in flour and protein concentrate of 1.87-4.47 μg TE/g and 0.82-3.62 μg TE/g, respectively. In addition, for the case of ferric reduction (FRAA) and phosphomolybdenum (AA), antioxidant activity methods showed values of 1.35-2.28 mg AAE/kg and 4262.7-5482.5 mg AAE/kg to flour, as well as 0.78-2.21 mg AAE/kg and 2229.1-3533.8 mg AAE/kg to protein concentrate, respectively.

## 2.2. Antimicrobial Activity

Plants, including their leaves, seeds, roots, and fruits, have been extensively studied for several years as potential sources of new natural compounds against emerging pathogens. Humans currently face fungal pathogens in food crop production and resistant microorganisms in multiple infections of clinical interest (Fisher et al., 2018). For this reason, many investigations focused on discovering antimicrobial compounds in plant-based foods, which are produced as a defense mechanism against biotic stress like microbial pathogens infection or insect attacks, as well as in response to abiotic stress such as drought, salinity, and other (Iqbal et al., 2021). Interesting compounds have been reported in bean seeds that could have antimicrobial activity, such as polyphenols, proteins, peptides, and different amylase inhibitors (Hamed et al., 2017; Ebrahim et al., 2022). However, few studies have focused on the bean's antimicrobial activity. We will describe some investigations that have been performed with this research approach.

Ebrahim et al. (2022) surveyed compounds from red kidney bean seeds (*Phaseolus vulgaris*), such as phenolic-rich extract, total protein extract, 7S globulin, and 11S globulin as antibacterial agents against multidrug-resistant (MDR) enterobacterales. Researchers documented that the total protein extract was the most effective, followed by 7S and 11S globulin, whereas the polyphenols extract was the least effective. In another study, isolated lectins from bean seeds were evaluated for their potential antibacterial activity against pathogens, observing that lectins were effective in inhibiting the growth of *Staphylococcus aureus* ATCC 6538, *Streptococcus mutants* ATCC 25175, *Pseudomonas aeruginosa* ATCC 10145, *Klebsiella pneumoniae,* and *Candida albicans* (Hamed et al., 2017). Other important discoveries regarding the antimicrobial compounds of beans have been observed in peptides encrypted in the storage proteins. For example, a protein hydrolysate from dark red kidney beans reveals antimicrobial activity against *Escherichia coli*, *P. aeruginosa*, *S. aureus*, and *Bacillus subtilis* (Roy et al., 2020). On the other hand, defensins-like peptides isolated from purple pole beans showed a potent antipathogenic effect against mycelial growth in *Mycosphaerella arachidicola*, *Helminthosporium maydis*, *Fusarium oxysporum*, *Verticillium dahliae*, *Rhizoctonia solani*, *C. albicans*, and *Setosphaeria turcica* (Lin et al., 2010). Similarly, Chan and Ng (2013) working with defensins-like peptides from the northeast red bean, found an antifungal activity against pathogenic fungi, including *Mycosphaerella arachidicola*, *Setosphaeria turcica*, *Fusarium oxysporum*, and *Valsa mali*. In addition, this peptide demonstrated remarkable thermostability and pH stability. Many studies have confirmed that encrypted peptides in the storage proteins from beans act as antimicrobial agents with potential biotechnological applications.

## 2.3. Anti-Obesity and Antidiabetic Activity

Concerns for better health in recent years have triggered a particular interest in deciphering various food components that promote health, with specific attention to diseases such as obesity, type 2 diabetes, and cardiovascular diseases. Obesity is a positive imbalance between energy intake and expenditure, with excessive weight gain linked to fat accumulation in adipose tissue. On the other hand, diabetes is described as a metabolic disorder in insulin concentrations or resistance to insulin actions, resulting in hyperglycemia. Both diseases are widely related, as well as other cardiovascular diseases (Castillo et al., 2019; Oussaada et al., 2019). Therefore, some investigations have focused on studying food components, revealing essential health benefits. Bean flour consumption has been reported to decrease the glycemic index and low-density lipoproteins (LDL) while increasing high-density lipoproteins (HDL) (Karaman et al., 2021).

The bean (*P. vulgaris*) has been recognized as one the most relevant food with anti-obesity and antidiabetic activity, demonstrating a meaningful efficacy against insulin resistance, fat accumulation, and dyslipidemia (Jakubczyk et al., 2017; Mojica et al., 2017a; Mojica et al., 2017b; Qin et al., 2019; Ashraf et al., 2020). Bioactive compounds, such as peptides and phenolic compounds, have been shown to have molecular targets with potential therapeutic effects on health. Several efforts have been made to decipher the mechanisms of action of bean bioactive compounds. Peptides derived from protein hydrolysates from enzymatic inhibition of dipeptidyl peptidase IV (DPP-IV), α-amylase, α-glucosidase, angiotensin I-converting enzyme (ACE), and lipases have been researched. In addition, efficacy assays have been done regarding cholesterol-lowering agents and glucose transporters blockade (Jakubczyk et al., 2017; Mojica et al., 2017a; Mojica et al., 2017b; Qin et al., 2019; Ashraf et al., 2020). Bean phenolic extracts show inhibitory activity against cholinesterases, amylase, glucosidase, and tyrosinase (Stefanucci et al., 2022).

Bean peptides can inhibit some enzymes linked to chronic diseases like type-2 diabetes, obesity, and hypertension. For instance, in the Brazilian Carioca and Mexican black beans, peptides such as CPGNK, KTYGL,

GGGLHK, and KKSSG with inhibition activity on α-glucosidase and DPP-IV have been documented (Mojica et al., 2017a). The black bean (Black-Otomi) hydrolyzed protein isolate was evaluated in glucose absorption in Wistar rats, revealing a reduction in postprandial glucose from 22.7% to 47.7% in diabetic rats. This trend was related to the blockade of the glucose transporters GLUT2 and SGLT1, evidence that bean peptides contribute to blood glucose control through hypoglycemic effects (Mojica et al., 2017b). In another study, pinto bean peptides (PPHMLP, PLPWGAGF, PPHMGGP, PLPLHMLP, and LSSLEMGSLGALFVCM) showed significant inhibitory ability on α-amylase, which could be allowing for the decreased glucose availability and, thus, acting to control diabetes (Ngoh et al., 2017). In the white kidney bean hydrolyzed proteins were reported capable of reducing cholesterol levels by inhibiting cholesterol solubilization and inhibiting cholesterol synthesis (Ashraf et al., 2020). In this study, using an *in vitro* cholesterol micelle model, it was found that protein hydrolysates from beans forced the substitution of cholesterol in the micelles, reducing its solubility and blocking its absorption in the human body. On the other hand, the authors studied the inhibition of the enzyme that participates in synthesizing cholesterol 3-hydroxy-3-methylglutaryl coenzyme A reductase (HMG-CoA), observing that the protein hydrolysates of the beans presented inhibition percentages above 60.

Interesting outcomes have been found in other studies using different approaches to promote peptide generation. In Pinto Durango and black 8025 beans, peptides like LLSV, WVVL, and ALVLL were reported from hard-to-cook bean proteins using different hydrolysis enzymes (alcalase, bromelain, pepsin, and pancreatin). Here an increase in insulin secretion, DPP-IV expression reduction, lipid accumulation inhibition, and glucose uptake increase were reported (Oseguera-Toledo, 2015). On the other hand, in fermented bean seeds with *Lactobacillus plantarum*, a peptide release was observed with the highest lipase inhibitory ($IC_{50}$ 1.19 mg/mL) and ACE (0.28 mg/mL) activity. In this peptide fraction, peptides such as INEGSLLLPH, GGYQGGGYGGNSGGGYGNRG, FVVAEQAGNEEGFE, GSGGGGG-GG, FGGPRR, SGGGGGGVAGAATASR, GGSGGGGGSSSGRRP, and GDTVTVEFDTFLSR were identified (Jakubczyk et al., 2017).

Another important result has been found in aqueous extracts of bean seeds. A study conducted with the extract of *P. vulgaris* using C57BL/6 mouse models fed a high-fat diet, a reduced body weight, blood glucose, triglyceride, and cholesterol levels were observed (Micheli et al., 2019). A green bean extract has shown lipolytic activity both in *in vitro* and *ex vivo* assays. Furthermore, bean extracts have shown an anti-adaptogenic effect during the differentiation of 3T3-L1 preadipocytes (Castillo et al., 2019). Some authors have suggested that the anti-obesity activity of bean extracts is associated with the inhibitor isoform I of α-amylase designated as phaseolamin. DeGouveia et al. (2014) demonstrated that phaseolamin can inhibit α-amylase activity and reduce blood glucose levels in streptozotocin-induced diabetic rats. Nonetheless, polyphenols have also been found in bean extracts that could contribute to anti-obesity activities. For example, Zheng et al. (2020) revealed that Mung beans have polyphenols with inhibition activity on α-amylase and α-glucosidase; in addition, polyphenols of Adzuki bean extracts reduce total hepatic lipids and triglycerides accumulation (Kitano-Okada et al., 2012).

## 2.4. Anti-Inflammatory, Antiproliferative and Anticarcinogenic Activity

The inflammatory response promotes the activation of transcriptional factors and proinflammatory cytokines. The interaction between immune system cells (natural killer (NK), T and B lymphocytes, dendritic cells, and macrophages) and non-immune cells activate proinflammatory mediators (chemokines and cytokines) and generate inflammatory signals that enable the organism to recognize, destroy, and eliminate foreign elements resulting in a successful acute inflammatory response. Otherwise, inappropriate expression of proinflammatory mediators leads to immune suppression, inhibition of insulin signaling, and chronic inflammatory conditions such as tissue degeneration, cardiovascular, neurodegenerative, and autoimmune diseases, and carcinogenesis (Teodoro, 2019, Nasef and Mehta, 2020; Sanchez-Chino et al., 2022). Several bioactive compounds in the diet exhibit antioxidant, anti-inflammatory, antiproliferative, and chemopreventive effects, among them are phytochemicals, as shown in Figure 1 (Finetti et al., 2020; Perez-Hernandez et al., 2021). Perez-Hernandez et al. (2021) demonstrated that enzymatic digestion of the common bean of different varieties and colors promotes the bioaccessibility of polyphenols with anti-inflammatory and proinflammatory properties in macrophages.

**Table 1.** Biological properties of bioactive compounds in *P. vulgaris*.

| Bioactivity | Compounds | Variety and Color | Model or Assay | Results | Reference |
|---|---|---|---|---|---|
| Antioxidant activity | Peptides | Red kidney beans | *In vitro* | High pressure (HP) treatment of the protein hydrolysate increased from 12.4% to 15.2%, and an alcalase hydrolysate at atmospheric pressure (24.7%) compared to the control. The combination of the methods (300 MPa/1% alcalase) obtained the maximum value (30.1%) of DPPH radical scavenging activity. | Al-Ruwaih et al. (2019) |
| | Polyphenols | White bean var. Fagiola di Venanzio | *In vitro* | In colorectal adenocarcinoma cell model (HT29 and HCT116), it was observed that the phenolic extract possesses anti-radical activity (DPPH) and inhibited the ability of interleukin 1-β (IL-1β) to promote ROS production at concentrations of 10 and 50 μg/mL, highlighting that chlorogenic acid acts mainly as a radical neutralizer. | Finetti et al. (2020) |
| | Peptides | White Kidney beans | *In vitro* | Hydrolysates treated with ultrasound showed higher antioxidant activity by DPPH (64.58%) and FRAP (25.43%). | Ashraf et al. (2020) |
| | Polyphenols | Tondino del Tavo and Cannellino Bio | *In vitro* | The extracts showed antioxidant activity by DPPH (6.65 mg TE/g) and ABTS (55.96 mg TE/g) and reducing power by CUPRAC and FRAP (42.94 mg TE/g). The presence of these compounds indicates that it is possible to give importance to a variety that is not very conventional. | Stefanucci et al. (2022) |
| Antiobesity activity | Polyphenols | Green bean | *In vitro* and *In vivo* | Evaluated methanol extracts in mature 3T3-L1 adipocytes and in rat white adipose tissue in an *in vivo* model, they observed that at concentrations of 800 and 1000 μg/mL they increased glycerol release significantly. The 1000 μg/mL extract reduced 47.1% of the intracellular triglyceride accumulation compared to the control. The green bean extract exhibited a lipolytic and anti-adipogenic effect *In vitro* by exerting a direct mechanism on adipocytes through lipolysis. | Castillo et al. (2019) |

**Table 1.** (Continued)

| Bioactivity | Compounds | Variety and Color | Model or Assay | Results | Reference |
|---|---|---|---|---|---|
| Antidiabetic activity | Alpha-amylase inhibitor isoform I (α-AI) with phytohemagglutinin | White kidney beans | *In vivo* | A comparison of metformin (100 mg/kg), atorvastatin (10 mg/kg), and dry bean extract in a metabolic syndrome model showed that the extract normalized glucose and insulin tolerance at a concentration of 500 mg/kg for 19 weeks. It reduced the pathologies of a metabolic syndrome model and decreased hepatic steatosis and lipid peroxidation in the liver. Protected the heart from oxidative alterations by increasing the expression of the detoxifying enzymes, catalase and glutathione reductase, normalizing the level of NADH dehydrogenase, and reducing the development of fatty streaks in the aorta muscle layers. | Micheli et al. (2019) |
| | Phaseolamin (α-AI) | White kidney beans | *In vivo* | In a murine model, authors evaluated α-amylase extracts for 90 days, observing that bean extracts (4 g/kg) with a polypeptide pattern of 36 and 45 kDa exhibited α-amylase inhibitory activity (2.157 U/g) and produced no treatment-related changes in body weight, absolute organ weight, relative organ weight, food consumption, blood chemistry, and hematology in rats. | Qin et al. (2019) |
| Anti-inflammatory activity | Polyphenols | White bean var. Fagiola di Venanzio | *In vitro* | The extract reduced the expression of the inflammatory markers COX-2, NF-kB, and ERK1/2 MAPK activation, and IL1-β promoted cell growth in colon cancer cell models (HT29 and HCT116) at a concentration of 50 g/mL. | Finetti et al. (2020) |
| | | Black bean var. Jamapa | *In vitro* | The extracts reduced IL-6, IL1-β, and iNOS mRNA levels by up to 20%. They exhibited a moderate anti-inflammatory effect because they decreased IL-6 expression by 18 and 20% at concentrations of 50 and 100 μg/mL of extract, respectively. The extract (100 μg/mL) reduced 16% IL1-β mRNA levels and iNOS target gene mRNA levels. | Perez-Hernandez et al. (2021) |

| Bioactivity | Compounds | Variety and Color | Model or Assay | Results | Reference |
|---|---|---|---|---|---|
| Anti-proliferative activity | Polyphenols | White bean var. Fagiola di Venanzio" | *In vitro* | Authors observed that the extract at concentrations of 5 to 100 μg/mL was able to reduce cell growth and proliferation induced by IL1 in colon cancer cells. The extract plays an important role in preventing the alteration of molecular processes that characterize antioxidants. | Finetti et al. (2020) |
| | | Red beans var. Marcela and Yellow bean var. Azufrado | *In vitro* | The extracts induced IL1-β, IL-6, and HO-1 mRNA expression, inducing pro- and anti-oxidative pathways.<br>Azufrado bean increased IL-6 mRNA levels by 50% when incubated with 50 μg/mL. | Perez-Hernandez et al. (2021) |

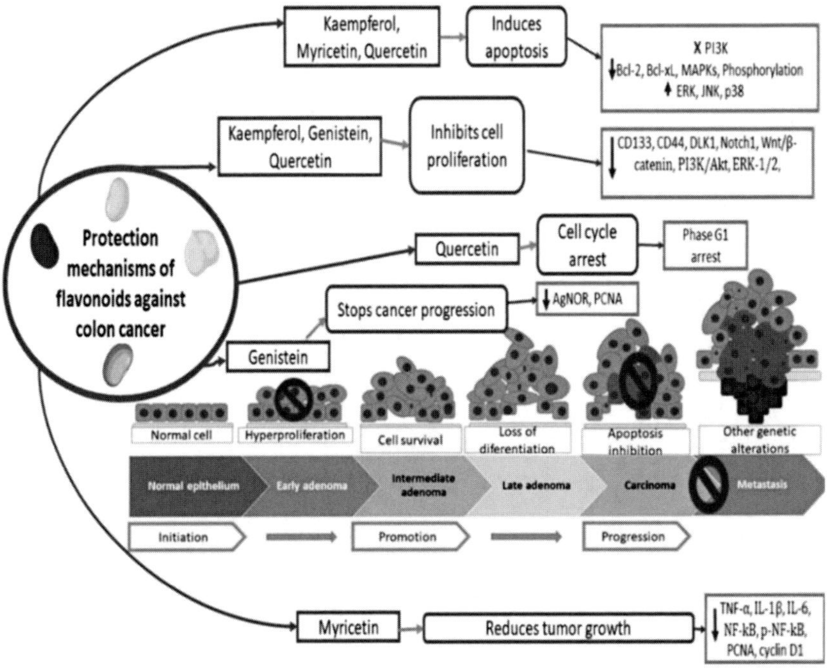


**Figure 1.** Protection mechanisms of flavonoids found in legumes on colon cancer (Sanchez-Chino et al., 2022).

## 3. Effect of Storage Processing on the Bioactivity of Beans

Bean consumption has increased due to its benefits, however, storage and processing play an important role in its chemical composition. Traditionally, beans are consumed after cooking, however, other processes such as extrusion, germination, fermentation, and roasting, which influence the sensory characteristics of the seeds, have been implemented. The changes that beans undergo during processing include the digestibility of their macronutrients, starch, and phytochemicals (Cárdenas-Castro et al., 2020; Martínez-Ávila et al., 2022; Perera et al., 2023).

### 3.1. Storage

Post-harvest storage of common beans is a strategy to ensure food security. The technological, physicochemical, and functional quality of beans is determined by the interaction between genotype, storage conditions, and high hygroscopicity. Seeds are usually stored in silos exposed to the elements, in regions with high temperatures. The heating of the silo surface can cause temperature gradients and water migration to the interior, so its high hygroscopicity causes an increase in its moisture content (Ferreira et al., 2018; Alves et al., 2020; Duarte and Badiale, 2022; Uebersax et al., 2022). Storage temperatures, use of improper packaging, relative humidity associated with moisture content of bean seeds, prolonged storage period, and presence of microorganisms constitute improper storage conditions (Ambachew et al., 2021; Cox et al., 2021; Perera et al., 2023).

Inside the bean seed, several reactions are activated during storage, among them the browning phenomenon, which is produced by several phenolic compounds, mainly condensed tannins (Alves et al., 2020; Nicolás-García et al., 2022), and these phenolic compounds are susceptible to different enzymatic and non-enzymatic browning reactions, mainly to their participation in oxidative stages. Nicolás-García et al. (2022) identified 37 phenolic compounds related to color saturation and increased antioxidant capacity after 9 months of storage (70% RH and 30 °C) that participate in the browning and hardening of black beans. On the other hand, Chen et al. (2022) observed that long-term storage caused a significant decrease in soluble phenolic compounds in the cotyledons. They identified abundant vanillin (bound phenolic compound) with a concentration of 2.38-27.25 μg/mg cellulose, 4-hydroxybenzoic acid (0.85-6.75 μg/mg cellulose), and 4-hydroxybenzaldehyde (0.18-2.43 μg/mg cellulose) in the cotyledon cell walls of a red kidney bean.

Water holding capacity is influenced by the cell wall composition of bean seeds. Inadequate storage has been reported to produce the phenomenon of hardening during shelf life caused by physical, structural, and chemical changes at the intercellular level, favoring the hard cooking defect (HTC) in beans (Chu et al., 2020; Nakhungu et al., 2020; Chen et al., 2022). A microstructural study by Nicolás-García et al. (2021c) revealed that the thickness of the common bean seed coat is 103 ± 9 mm, representing 10% of the whole seed, this coat protects the internal structure from mechanical injury, participates in gas exchange, and water uptake under natural conditions and during storage.

The phytate-phytate-pectin theory (Figure 2) suggests that storage at high temperatures and relative humidity increases phytase activity, leading to phytate dephosphorylation and reduced ability to chelate divalent cations ($Ca^{2+}$, $Fe^{2+}$, $Zn^{2+}$, and $Mg^{2+}$). These divalent cations migrate to the middle lamella of the cell wall of parenchymal cells, where they bind to the carboxyl groups of pectin, forming insoluble pectates, thus hardening the cell walls and originating the hardening phenomenon, which favors hard-to-cook "HTC" and reduces nutritional quality (Cominelli et al., 2020; Kyomugasho et al., 2021; Chen et al., 2023). Chen et al. (2022) reported that phenolic acids are strongly bound to the cell wall of aged beans compared to the fresh sample. They observed that phenolic compounds in the cell wall after aging interact with pectic polysaccharides, and structural carbohydrates such as hemicellulose and cellulose, in a polymerized form similar to lignin, evidencing the hardening process of the seed.

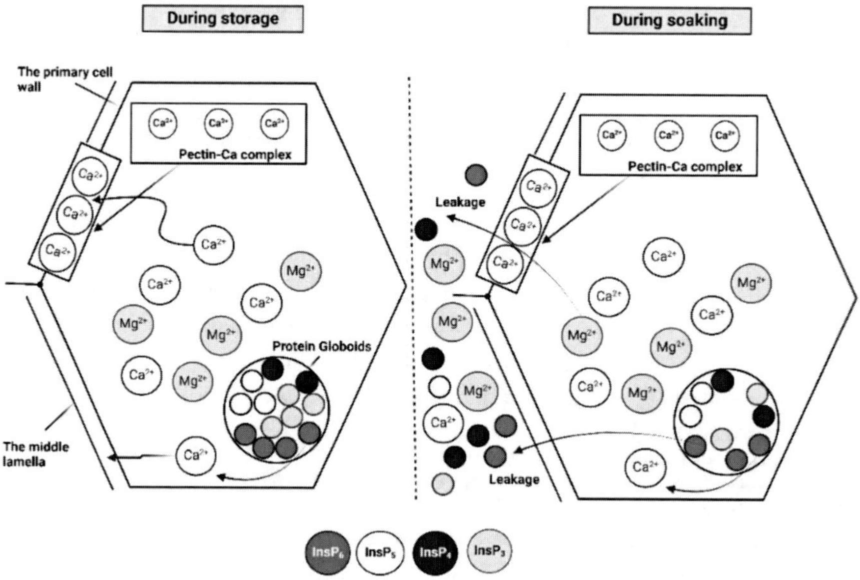


**Figure 2.** Pectin-cation-phytate mechanism of HTC development during postharvest storage and subsequent soaking of common bean (Chen et al., 2023). Created in Biorender.

## 3.2. Cooking

Currently, all commercial bean varieties are subjected to a cooking process before consumption, this treatment allows for increasing starch digestibility and modifies nutritional components such as proteins and polyphenols (Yang et al., 2018; Rebollo-Hernanz et al., 2019; Nakhungu et al., 2020). Inadequate storage conditions lead to the occurrence of the cooking hardness phenomenon (HTC), which, in turn, results in an increase in cooking time and a decrease in the nutritional and sensory quality of the common bean. During cooking, there are interactions or bonds caused by the solubilization of compounds present in the bean seed coat, this includes divalent cationic cross-linkages between demethoxylated pectic substances and phenolic cross-linkages (Chigwedere et al., 2019; Wainaina et al., 2021).

Alteration of the cell wall matrix of legume seeds by heat application can improve starch and protein digestibility by increasing their exposure and making them susceptible to enzymatic hydrolysis (Alpos et al., 2021). Studies conducted so far related to the effect of the cooking process on beans have mainly focused on the study of the endosperm, and starch granules in hardened beans (Mujica et al., 2015) or grown under different water regimes (Miano et al., 2018). Textural changes during thermal processing of high-starch foods involve hydration, solubilization, swelling, and loss of the crystalline structure of starch granules, dissolution of the cell wall and lamellar polymers found in the cotyledon, leading to a breakdown in the cell walls, facilitating the release of nutrients from the food matrix and, at the same time, causing a change in coloration (Chigwedere et al., 2018).

Heat treatment can improve protein digestibility by destroying protease inhibitors, opening the protein structure through denaturation and elimination of soluble solids. Total dietary fiber content increases during cooking due to the formation of resistant starch or the formation of tannin-protein complexes (Corzo-Ríos et al., 2020). According to the literature, the cooking process also reduces the phytate and oxalate content in seeds, which reduces the bioavailability of calcium, zinc, and iron (Mwanri et al., 2018). Specific phenolic compounds such as ferulic acid, sinapic acid, and bound flavanols increase in concentration after the cooking process (Chen et al. 2019).

Cárdenas-Castro et al. (2020) evaluated the bioaccessibility and *in vitro* release kinetics of phenolic compounds from two varieties of beans (Azufrado and Negro Jamapa). The authors reported that in cooked beans the phenolic compounds showed 50% bioaccessibility and 30% in cooked-fried beans, indicating that cooking did not modify the release kinetics of phenolic compounds during the first 60 min, identifying kaempferol-3-*O*-glucoside, quercetin-3-*O*-glucoside, and chlorogenic acid as the main compounds released. In addition, Perez-Hernandez et al. (2021) observed that the cooking process (boiling 70 min; 1:10, w: v) caused a decrease in total phenol content in the common bean variety Marcela (red seeds) with a 74% decrease, in contrast to the black Jamapa variety increased 55% compared to the raw sample. The red beans showed a decrease in antioxidant activity compared to the black bean, the latter doubling the activity concerning the control in the TEAC and FRAP assays.

Cooking common bean seeds results in new sensory properties such as flavor, aroma, and texture of the seed (Teixeira-Guedes et al., 2019), improving bioaccessibility during digestion and availability of diols for beneficial health effects (Cárdenas-Castro et al., 2020; Wainaina et al., 2021). Gomes et al. (2021) observed in an *in vivo* model that consumption of cooked beans and protein concentrate increased the abundance of Bacteroidetes and decreased Firmicutes/Bacteroidetes compared to the control, showing that consumption of cooked beans improved intestinal health and modulated the composition and function of the cecal microbiota, attenuating the effects of a high-fat diet, which has been attributed to the beans' chemical composition and high fiber content.

## 3.3. Germination

During seed germination of some legumes, there is a degradation of reserve proteins that corresponds to an accumulation of free amino acids in the cotyledons. Digestion of the protein bodies begins at the periphery and progressively advances towards the center of the cotyledons. The maximum proteolytic activity occurs 5 days after germination, which coincides with the maximum accumulation of free amino acids. Therefore, germination can cause significant changes in the biochemical characteristics of seeds and, during this process, proteases can degrade storage proteins, increasing their nutritional value and improving protein digestibility, reducing antinutritional factors and hydrolyzing oligosaccharides (Cárdenas-Castro et al., 2020; Muñoz-Llandes et al., 2022, Bera et al. 2023).

Germination begins when the dry seed starts to absorb water and is completed when the embryonic axis elongates. From the moment the seed breaks dormancy, protective responses arise through the synthesis of phenols and other compounds (Farooq et al., 2019). In addition to seeds, germinated seed sprouts are also served as a staple vegetable in many countries. Germination can lead to changes in nutrients, including functional sub-states, and can remove anti-nutrients such as enzyme inhibitors from the seeds (Yang et al., 2018; Muñoz-Llandes et al., 2022; Bera et al., 2023). Rossi et al. (2022) evaluated the impact of germination time and gastrointestinal digestion of *P.*

*vulgaris* by observing a relationship between phenolic content and soil composition.

In dicotyledonous seeds, storage proteins such as legumins and vicilins are degraded during germination to provide the necessary nutrition and seedling growth (Bera et al., 2023). During germination, proteins begin to degrade by the action of endopeptidases located in the cotyledons of the seeds forming oligopeptides. The oligopeptides are hydrolyzed by exopeptidases to form free amino acids, which will be used to synthesize new proteins and tissues (Borges-Martínez et al., 2021). In addition, the germination process of common beans has been reported to impact total phenol content, increasing after 55 days of germination (Stefanucci et al., 2022).

## 3.4. Roasting

Roasting is a very popular heating process in Eastern countries, characterized by changing the physical (color) and chemical properties of bean seeds. Several studies have shown that antioxidant activity increases after certain heat treatments due to the Maillard reaction as it breaks cell walls and membranes, which releases insoluble ester bonds to soluble phenolic substances that may also contribute to increase the antioxidant property (Roy et al., 2021; Alpos et al., 2021; Gancarz et al., 2022). Bean seeds are rich in phenolic compounds ranging from simple molecules such as phenolic acids and flavonoids to polymerized compounds such as lignins (Król et al., 2020; Nicolás-García et al., 2021b).

In the roasting process, heat can be transferred to the beans by three methods: convection, conduction, or radiation. The most important variables for each type of roasting are the type of roaster, the temperature of the beans, and the duration of roasting (Fadri et al., 2021). For example, convection heat transfer is based on hot air rising from the hot surfaces towards the coffee beans, whereas conduction heat is transferred through direct contact of the coffee beans with the hot surfaces of the rotating drum (Gancarz et al., 2022). Roy et al. (2021) observed that the phenolic content in dark red kidney bean decreased (8.15 mg GAE/g) after roasting compared to the raw sample (12.55 mg GAE/g).

The increase in phenolic content depends mainly on the type of legume and the preparation procedure used, with roasting being the method that preserves the highest content of phenolic compounds and flavonoids in dry seeds, compared to boiling or cooking. Prior to the roasting process, bean seeds are usually soaked, which helps to soften the texture and shortens the cooking time. However, this soaking process can increase the permeability of the seed coat, and the thermal process can cause degradation of phenolics by leaching (Zhou et al., 2017; Gancarz et al., 2022).

## 3.5. Extrusion

Extrusion cooking is one of the most important food processing technologies for obtaining different foods, it improves digestibility and nutrient bioavailability compared to conventional cooking and has textural advantages, such as expansion, crispness, low operating costs, energy efficiency, and shorter cooking times. Extrusion is a high temperature short time (HTST) process with the purpose of preserving essential nutrients, reducing non-nutritional components and microorganisms. It has been reported that during the extrusion process antioxidant properties increase (Escobedo et al., 2020; Félix-Medina et al., 2020; Cotacallapa-Sucapuca et al., 2021). Extrusion is an effective method to improve the digestibility of protein and starch from extruded bean seeds. Resistant starch is a fraction of starch that resists digestion, reaches the colon to be fermented by the microbiota, producing short-chain fatty acids such as acetate, propionate, and butyrate that contribute to the prevention of colon cancer. Several studies have used the extrusion method to obtain type 3 resistant starch (retrograded), which alters the starch granules by increasing the retrogradation process (Ramírez-Jiménez et al., 2018; Escobedo et al., 2020).

The factors influencing the extrusion of bean seeds include various extrusion conditions such as barrel temperature, screw speed, pressure, shear force, extrusion temperature, moisture content, duration, and pH. These parameters significantly impact the functional properties of the proteins, affecting their structure, solubility, and bioactivity during the extrusion process. Under the synergistic action of high temperature, pressure, and shear in the extruder, the protein structure will unfold, linearize, and cross-link, which will weaken the binding strength to maintain the tertiary and quaternary structures of the protein. At that moment, the amino acids of the protein will reorganize and change from a spherical to a fibrous aggregation, the spatial conformation of the protein will change, and its denaturation will occur (Félix-Medina et al., 2020; Yang et al., 2020; Cotacallapa-Sucapuca et al., 2021). Extrusion enhances the release of bound phenolic compounds, which influences the sensory properties of the seeds. It has been reported that the anthocyanin content of extruded beans decreases, whereas flavonols increase in concentration (Nicolás-García et al., 2021b). Félix-Medina et al. (2020) observed that the extrusion process to produce snacks incorporated with 16 to 40% bean flour caused a decrease in lipid content and an increase in ash and carbohydrate content, mainly dietary fiber. The content of total phenols and antioxidant capacity (ABTS, DPPH, and FRAP) increased significantly, mainly the bound phenolic fraction followed by the free phenolic fraction. The authors mention that this may be due to the conditions of humidity, temperature, and speed in the process.

## Conclusion

The biodiversity of food resources is vast, enabling the provision of nutrient-rich foods such as legumes, which serve as valuable sources of healthy nutrients for consumers. Among these, the common bean stands out as a staple food with significant genetic diversity and numerous varieties. Beans provide essential nutrients, including proteins, carbohydrates, lipids, vitamins, and minerals, as well as bioactive components like saponins and phenolic compounds. However, during processing for consumption, aimed at ensuring food safety, these components undergo modifications that enhance their bioavailability, making them more beneficial to the body. Research on bioactive compounds has demonstrated the biological properties present in beans, which contribute to promoting and maintaining health stability in consumers.

## References

Alcázar-Valle, M., García-Morales, S., Mojica, L., Morales-Hernández, N., Sánchez-Osorio, E., Flores-López, L., Enríquez-Vara, J. N., & Lugo-Cervantes, E. (2021). Nutritional, antinutritional compounds and nutraceutical significance of native bean species (*Phaseolus* spp.) of Mexican cultivars. *Agriculture*, *11* (11), 1031. doi:10.3390/agriculture11111031.

Alcázar-Valle, M., Lugo-Cervantes, E., Mojica, L., Morales-Hernández, N., Reyes-Ramírez, H., Enríquez-Vara, J. N., & García-Morales, S. (2020). Bioactive compounds, antioxidant activity, and antinutritional content of legumes: a comparison between four *Phaseolus* species. *Molecules*, *25* (15), 3528. doi:10.3390/molecules25153528.

Alpos, M., Leong, S. Y., & Oey, I. (2021). Combined effects of calcium addition and thermal processing on the texture and *in vitro* digestibility of starch and protein of black beans (*Phaseolus vulgaris*). *Foods*, *10* (6), 1368. doi:10.3390/foods10061368.

Al-Ruwaih, N., Ahmed, J., Mulla, M. F., & Arfat, Y. A. (2019). High-pressure assisted enzymatic proteolysis of kidney beans protein isolates and characterization of hydrolysates by functional, structural, rheological and antioxidant properties. *LWT*, *100*, 231-236. doi:10.1016/j.lwt.2018.10.074.

Alves, N. E., Gomes, M. J., Vasconcelos, C. M., Lima, A. C., de Lima, S. L., Brito, E. S., ... & Martino, H. S. (2021). Six months under uncontrolled relative humidity and room temperature changes technological characteristics and maintains the physicochemical and functional properties of carioca beans (*Phaseolus vulgaris* L.). *Food Chemistry*, 342, 128390. doi:10.1016/j.foodchem.2020.128390.

Ambachew, D., Joshua, J., Mmbaga, M. T., & Blair, M. W. (2021). Sources of resistance to common bacterial blight and charcoal rot disease for the production of mesoamerican common beans in the southern United States. *Plants*, *10* (5), 998. doi:10.3390/plants10050998.

Aravindakshan, S., Kyomugasho, C., Tafiire, H., Van Loey, A., Grauwet, T., & Hendrickx, M. E. (2022). The moisture plasticizing effect on enzyme-catalyzed reactions in model and real systems in view of legume ageing and their hard to cook development. *Journal of Food Engineering*, *314*, 110781. doi:10.1016/j.jfoodeng.2021.110781.

Ashraf, J., Liu, L., Awais, M., Xiao, T., Wang, L., Zhou, X., Tong, L.T, & Zhou, S. (2020). Effect of thermosonication pre-treatment on mung bean (*Vigna radiata*) and white kidney bean (*Phaseolus vulgaris*) proteins: Enzymatic hydrolysis, cholesterol lowering activity and structural characterization. *Ultrasonics Sonochemistry*, *66*, 105121. doi:10.1016/j.ultsonch.2020.105121.

Bera, I., O'Sullivan, M., Flynn, D., & Shields, D. C. (2023). Relationship between protein digestibility and the proteolysis of legume proteins during seed germination. *Molecules*, *28* (7), 3204. doi:10.3390/molecules28073204.

Capistrán-Carabarin, A., Aquino-Bolaños, E. N., García-Díaz, Y. D., Chávez-Servia, J. L., Vera-Guzmán, A. M., & Carrillo-Rodríguez, J. C. (2019). Complementarity in phenolic compounds and the antioxidant activities of *Phaseolus coccineus* L. and *P. vulgaris* L. Landraces. *Foods*, *8* (8), 295.doi:10.3390/foods8080295.

Cárdenas-Castro, A. P., Pérez-Jiménez, J., Bello-Pérez, L. A., Tovar, J., & Sáyago-Ayerdi, S. G. (2020). Bioaccessibility of phenolic compounds in common beans (*Phaseolus vulgaris* L.) after *in vitro* gastrointestinal digestion: A comparison of two cooking procedures. *Cereal Chemistry*, *97* (3), 670–680. doi:10.1002/cche.10283.

Castillo, F., González, D. R., & Moore-Carrasco, R. (2019). Effects of *Phaseolus vulgaris* extract on lipolytic activity and differentiation of 3T3-L1 Preadipocytes into mature adipocytes: A strategy to prevent obesity. *Journal of Nutrition and Metabolism*, *2019*. doi:10.1155/2019/5093654.

Chan, Y. S., & Ng, T. B. (2013). Northeast red beans produce a thermostable and pH-stable defensin-like peptide with potent antifungal activity. *Cell Biochemistry and Biophysics, 66*, 637-648. doi:10.1007/s12013-012-9508-1.

Chen, D., Ding, A., Zhu, L., Grauwet, T., Van Loey, A., Hendrickx, M., & Kyomugasho, C. (2023). Phytate and mineral profile evolutions to explain the textural hardening of common beans (*Phaseolus vulgaris* L.) during postharvest storage and soaking: Insights obtained through a texture-based classification approach. *Food Chemistry*, *404*, 134531. doi:10.1016/j.foodchem.2022.134531.

Chen, D., Hu, K., Zhu, L., Hendrickx, M., & Kyomugasho, C. (2022). Cell wall polysaccharide changes and involvement of phenolic compounds in ageing of Red haricot beans (*Phaseolus vulgaris*) during postharvest storage. *Food Research International*, *162*, 112021. doi:10.1016/j.foodres.2022.112021.

Chen, Z., Li, W., Santhanam, R. K., Wang, C., Gao, X., Chen, Y., Wang, C., Xu, L., & Chen, H. (2019). Bioactive peptide with antioxidant and anticancer activities from black soybean [*Glycine max* (L.) Merr.] byproduct: isolation, identification and molecular docking study. *European Food Research and Technology*, *245* (3), 677-689. doi:10.1007/s00217-018-3190-5.

Chigwedere, C. M., Nkonkola, C. M., Rai, S., Kyomugasho, C., Kermani, Z. J., Pallares, A. P., & Hendrickx, M. E. (2019). Cotyledon pectin molecular interconversions explain pectin solubilization during cooking of common beans (*Phaseolus vulgaris*). *Food Research International*, *116*, 462-470. doi:10.1016/j.foodres.2018.08.062.

Chigwedere, C. M., Olaoye, T. F., Kyomugasho, C., Jamsazzadeh Kermani, Z., Pallares Pallares, A., Van Loey, A. M., Grauwet, T., & Hendrickx, M. E. (2018). Mechanistic insight into softening of Canadian wonder common beans (*Phaseolus vulgaris*) during cooking. *Food Research International*, *106*, 522–531. doi:10.1016/j.foodres. 2018.01.016.

Chu, J., Ho, P., & Orfila, C. (2020). Growth region impacts cell wall properties and hard-to-cook phenotype of canned navy beans (*Phaseolus vulgaris*). *Food and Bioprocess Technology*, 1-9.doi:10.1007/s11947-020-02436-7.

Cominelli, E., Galimberti, M., Pongrac, P., Landoni, M., Losa, A., Paolo, D. Daminati M. G., Bollini, R., Cichy, K. A., Voguel-Mikus, K., & Sparvoli, F. (2020). Calcium redistribution contributes to the hard-to-cook phenotype and increases PHA-L lectin thermal stability in common bean low phytic acid 1 mutant seeds. *Food Chemistry*, *321*, 126680. doi:10.1016/j.foodchem.2020.126680.

Corzo-Ríos, L. J., Sánchez-Chino, X. M., Cardador-Martínez, A., Martínez-Herrera, J., & Jiménez-Martínez, C. (2020). Effect of cooking on nutritional and non-nutritional compounds in two species of *Phaseolus* (*P. vulgaris* and *P. coccineus*) cultivated in Mexico. *International Journal of Gastronomy and Food Science*, *20*, 100206. doi:10.1016/j.ijgfs.2020.100206.

Cotacallapa-Sucapuca, M., Vega, E. N., Maieves, H. A., Berrios, J. D. J., Morales, P., Fernández-Ruiz, V., & Cámara, M. (2021). Extrusion process as an alternative to improve pulses products consumption. A review. *Foods*, *10* (5), 1096. doi:10.3390/foods10051096.

Cox, L. D., Munholland, S., Mats, L., Zhu, H., Crosby, W. L., Lukens, L., Pauls, K. P., & Bozzo, G. G. (2021). The induction of the isoflavone biosynthesis pathway is associated with resistance to common bacterial blight in *Phaseolus vulgaris* L. *Metabolites*, *11* (7), 433. doi:10.3390/metabo11070433.

Duarte Santos, T., & Badiale Furlong, E. (2022). Biological contamination of the common bean (*Phaseolus vulgaris* L.) and its impact on food safety. *Critical Reviews in Food Science and Nutrition*, *62* (18), 4998-5004. doi:10.1080/10408398.2021.1881038.

Ebrahim, A. E., Abd El-Aziz, N. K., Elariny, E. Y., Shindia, A., Osman, A., Hozzein, W. N., & El-Hossary, D. (2022). Antibacterial activity of bioactive compounds extracted from red kidney bean (*Phaseolus vulgaris* L.) seeds against multidrug-resistant Enterobacterales. *Frontiers in Microbiology*, *13*. doi:10.3389/fmicb.2022.1035586.

Escobedo, A., Loarca-Piña, G., Gaytan-Martínez, M., Orozco-Avila, I., & Mojica, L. (2020). Autoclaving and extrusion improve the functional properties and chemical composition of black bean carbohydrate extracts. *Journal of Food Science*, *85* (9), 2783-2791. doi:10.1111/1750-3841.15356.

Fadri, R. A., Sayuti, K., Nazir, N., & Suliansyah, I. (2021). Sensory Quality Profile of Ranah Minang Arabica Coffee Specialty. *International Journal on Advanced Science Engineering Information Technology*, *11*, 281-290.

Farooq, M., Usman, M., Nadeem, F., ur Rehman, H., Wahid, A., Basra, S. M., & Siddique, K. H. (2019). Seed priming in field crops: Potential benefits, adoption and challenges. *Crop and Pasture Science*, *70* (9), 731-771. doi:10.1071/CP18604.

Félix-Medina, J. V., Montes-Ávila, J., Reyes-Moreno, C., Perales-Sánchez, J. X. K., Gómez-Favela, M. A., Aguilar-Palazuelos, E., & Gutiérrez-Dorado, R. (2020). Second-generation snacks with high nutritional and antioxidant value produced by an optimized extrusion process from corn/common bean flours mixtures. *LWT*, *124*, 109172. doi:10.1016/j.lwt.2020.109172.

Ferreira, C. D., Ziegler, V., da Silva Lindemann, I., Hoffmann, J. F., Vanier, N. L., & de Oliveira, M. (2018). Quality of black beans as a function of long-term storage and moldy development: Chemical and functional properties of flour and isolated protein. *Food Chemistry*, *246*, 473-480. doi:10.1016/j.foodchem.2017.11.118.

Finetti, F., Biagi, M., Ercoli, J., Macrì, G., Miraldi, E., & Trabalzini, L. (2020). *Phaseolus vulgaris* L. var. venanzio grown in Tuscany: Chemical composition and *in vitro* investigation of potential effects on colorectal cancer. *Antioxidants*, *9* (12), 1181. doi:10.3390/antiox9121181.

Fisher, M. C., Hawkins, N. J., Sanglard, D., & Gurr, S. J. (2018). Worldwide emergence of resistance to antifungal drugs challenges human health and food security. *Science*, *360* (6390), 739-742. doi: 10.1126/science.aap7999.

Gancarz, M., Dobrzański Jr, B., Malaga-Toboła, U., Tabor, S., Combrzyński, M., Ćwikła, D., & Rusinek, R. (2022). Impact of coffee bean roasting on the content of pyridines determined by analysis of volatile organic compounds. *Molecules*, *27* (5), 1559. doi:10.3390/molecules27051559.

Gomes, M. J. C., da Silva, J. S., Assis, A., de Mejia, E. G., Mantovani, H. C., & Martino, H. S. D. (2021). Common bean (*Phaseolus vulgaris* L.) flour can improve the gut microbiota composition and function in mice fed a high-fat diet. *Current Developments in Nutrition*, *5*, 5141159. doi:10.1093/cdn/nzab054_014.

Hamed, E. S., Ibrahim, E. A., & Mounir, S. M. (2017). Antimicrobial Activities of Lectins Extracted from Some Cultivars of *Phaseolus vulgaris* Seeds. *Journal of Microbial & Biochemical Technology*, *9* (3), 109-116. doi: 10.4172/1948-5948.1000352.

Iqbal, Z., Iqbal, M. S., Hashem, A., Abd_Allah, E. F., & Ansari, M. I. (2021). Plant defense responses to biotic stress and its interplay with fluctuating dark/light conditions. *Frontiers in Plant Science*, *12*, 631810. doi:10.3389/fpls.2021.631810.

Jakubczyk, A., Karaś, M., Złotek, U., & Szymanowska, U. (2017). Identification of potential inhibitory peptides of enzymes involved in the metabolic syndrome obtained by simulated gastrointestinal digestion of fermented bean (*Phaseolus vulgaris* L.) seeds. *Food Research International*, *100*, 489-496. doi:10.1016/j.foodres. 2017.07.046.

Karaman, K., Kaplan, M., Çiftci, B., & Yürürdurmaz, C. (2021). Color based genotypic differences effect the techno-functional and bioactive properties of common bean (*Phaseoulus vulgaris* L.) protein concentrates. *Research Square, 1-13.* doi:10.21203/rs.3.rs-478103/v1.

Kitano-Okada, T., Ito, A., Koide, A., Nakamura, Y., Han, K. H., Shimada, K. Sasaki, K., Ohba, K., Sibayama, S., & Fukushima, M. (2012). Anti-obesity role of adzuki bean extract containing polyphenols: *in vivo* and *in vitro* effects. *Journal of the Science of Food and Agriculture*, *92* (13), 2644-2651. doi:10.1002/jsfa.5680.

Król, K., Gantner, M., Tatarak, A., & Hallmann, E. (2020). The content of polyphenols in coffee beans as roasting, origin and storage effect. European *Food Research and Technology*, *246*, 33-39. doi:10.1007/s00217-019-03388-9.

Kyomugasho, C., Kamau, P. G., Aravindakshan, S., & Hendrickx, M. E. (2021). Evaluation of storage stability of low moisture whole common beans and their fractions through the use of state diagrams. *Food Research International*, *140*, 109794. doi:10.1016/j.foodres.2020.109794.

Lin, P., Wong, J. H., & Ng, T. B. (2010). A defensin with highly potent antipathogenic activities from the seeds of purple pole bean. *Bioscience Reports*, *30* (2), 101-109. doi:10.1042/BSR20090004.

Borges-Martínez, J. E., Moguel-Concha D. R., Gallardo-Velázquez, T. G., Jiménez-Martínez C., & Ruiz-Ruiz, J. C. (2021). Anti-inflammatory properties of phenolic extracts from *Phaseolus vulgaris* and *Pisum sativum* during germination. *Food Bioscience*, *42*, 101067. doi:10.1016/j.fbio.2021.101067.

Martínez-Ávila, M., Rodríguez-Rodríguez, J., Uribe, J. G., & Guajardo-Flores, D. (2022). Selective supercritical fluid extraction of non-polar phytochemicals from black beans (*Phaseolus vulgaris* L.) by-products. *The Journal of Supercritical Fluids*, *189*, 105730. doi:10.1016/j.supflu.2022.105730.

Mecha, E., Leitão, S. T., Carbas, B., Serra, A. T., Moreira, P. M., Veloso, M. M., & Bronze, M. R. (2019). Characterization of soaking process' impact in common beans phenolic composition: Contribute from the unexplored Portuguese germplasm. *Foods*, *8* (8), 296. doi:10.3390/foods8080296.

Miano, A. C., Saldaña, E., Campestrini, L. H., Chiorato, A. F., & Augusto, P. E. D. (2018). Correlating the properties of different carioca bean cultivars (*Phaseolus vulgaris*) with their hydration kinetics. *Food Research International*, *107*, 182–194. doi:10.1016/j.foodres.2018.02.030.

Micheli, L., Lucarini, E., Trallori, E., Avagliano, C., De Caro, C., Russo, R., Calignano, A., Ghelardini, C., Pacini, A., & Di Cesare Mannelli, L. (2019). *Phaseolus vulgaris* L. extract: Alpha-amylase inhibition against metabolic syndrome in mice. *Nutrients*, *11* (8), 1778. doi:10.3390/nu11081778.

Mojica, L., de Mejia, E. G., Granados-Silvestre, M. Á., & Menjivar, M. (2017a). Evaluation of the hypoglycemic potential of a black bean hydrolyzed protein isolate and its pure peptides using *in silico, in vitro* and *in vivo* approaches. *Journal of Functional Foods, 31*, 274-286. doi:10.1016/j.jff.2017.02.006.

Mojica, L., Luna-Vital, D. A., & González de Mejía, E. (2017b). Characterization of peptides from common bean protein isolates and their potential to inhibit markers of type-2 diabetes, hypertension and oxidative stress. *Journal of the Science of Food and Agriculture, 97* (8), 2401-2410. doi:10.1002/jsfa.8053.

Mujica, M. V., Granito, M., & Soto, N. (2015). Cambios microestructurales en los granos de *Phaseolus vulgaris* endurecidos. *Archivos Latinoamericanos de Nutrición, 65* (2), 110–118.

Muñoz-Llandes, C. B., Guzmán-Ortiz, F. A., Román-Gutiérrez, A. D., Palma-Rodríguez, H. M., Castro-Rosas, J., Hernández-Sánchez, H., & Vargas-Torres, A. (2022). Effect of germination time on protein subunits of *Lupinus angustifolius* L. and its influence on functional properties and protein digestibility. *Food Science and Technology, 42*. doi:10.1590/fst.90821.

Mwanri, A. W., Mamboleo, T. F., Msuya, J. M., & Gowele, V. F. (2018). Oxalate, phytate and nitrate content in African nightshade, spider plant and amaranths at different stages of maturity. *African Journal of Food Science, 12* (11), 316–322. doi:10.5897/AJFS2018.1735.

Nakhungu, W. E., Njoki, W. I., Carolien, B., Nghia-Do-Trong, N., Kahenya, K. P., Wouter, S., Marc, H. (2020). Application of near-infrared spectroscopy to predict the cooking times of aged common beans (*Phaseolus vulgaris* L.). *Journal of Food Engineering, 284*, 110056. doi:10.1016/j.jfoodeng.2020.110056.

Nasef, N. A., & Mehta, S. (2020). Role of inflammation in pathophysiology of colonic disease: An update. *International Journal of Molecular Sciences, 21* (13), 4748. doi:10.3390/ijms21134748.

Ngoh, Y. Y., Choi, S. B., & Gan, C. Y. (2017). The potential roles of Pinto bean (*Phaseolus vulgaris* cv. Pinto) bioactive peptides in regulating physiological functions: Protease activating, lipase inhibiting and bile acid binding activities. *Journal of Functional Foods, 33*, 67-75. doi:10.1016/j.jff.2017.03.029.

Nicolás-García, M., Perucini-Avendaño, M., Jiménez-Martínez, C., Perea-Flores, M. D. J., Gómez-Patiño, M. B., Arrieta-Báez, & D., Dávila-Ortiz, G. (2021a). Bean phenolic compound changes during processing: Chemical interactions and identification. *Journal of Food Science, 86* (3), 643-655. doi:10.1111/1750-3841.15632.

Nicolás-García, M., Jiménez-Martínez, C., Perucini-Avendaño, M., Camacho-Díaz, B. H., Jiménez-Aparicio, A. R., & Dávila-Ortiz, G. (2021b). Phenolic compounds in legumes: composition, processing and gut health by Jimenez-Lopez J. C. (Ed). IntechOpen. ISBN 978-1-83969-491-2. doi:10.5772/intechopen.98202.

Nicolás-García, M., Arzate-Vázquez, I., de Jesús Perea-Flores, M., Méndez-Méndez, J. V., Perucini-Avendaño, M., Gómez-Patiño, M. B., & Dávila-Ortiz, G. (2021c). An overview of instrumented indentation technique for the study of micromechanical properties in food: A case study on bean seed coat. *Biosystems Engineering*, 204, 377-385. doi:10.1016/j.biosystemseng.2021.02.006.

Nicolás-García, M., Perucini-Avendaño, M., Arrieta-Báez, D., de Jesús Perea-Flores, M., Jiménez-Martínez, C., Gómez-Patiño, M. B., & Dávila-Ortiz, G. (2022). Phenolic compounds profile by UPLC-ESI-MS in black beans and its distribution in the seed coat during storage. *Food Chemistry*, *395*, 133638. doi:10.1016/j.foodchem. 2022.133638.

Oseguera-Toledo, M. E., de Mejia, E. G., & Amaya-Llano, S. L. (2015). Hard-to-cook bean (*Phaseolus vulgaris* L.) proteins hydrolyzed by alcalase and bromelain produced bioactive peptide fractions that inhibit targets of type-2 diabetes and oxidative stress. *Food Research International*, *76*, 839-851. doi:10.1016/j.foodres.2015.07.046.

Oussaada, S. M., van Galen, K. A., Cooiman, M. I., Kleinendorst, L., Hazebroek, E. J., van Haelst, M. M., Ter Horst, K. W., & Serlie, M. J. (2019). The pathogenesis of obesity. *Metabolism*, *92*, 26–36. doi:10.1016/j.metabol.2018.12.012.

Perera, D., Devkota, L., Garnier, G., Panozzo, J., & Dhital, S. (2023). Hard-to-cook phenomenon in common legumes: chemistry, mechanisms and utilisation. *Food Chemistry*, 135743. doi:10.1016/j.foodchem.2023.135743.

Perez-Hernandez, L. M., Hernández-Álvarez, A. J., Morgan, M., Boesch, C., & Orfila, C. (2021). Polyphenol bioaccessibility and anti-inflammatory activity of Mexican common beans (*Phaseolus vulgaris* L.) with diverse seed colour. *CyTA-Journal of Food*, *19* (1), 682-690. doi:10.1080/19476337.2021.1965660.

Qin, G., Wang, F., Liang, H., Tang, S., Shekh, K., Wang, Y., Li, B., Dong, B., & Wen, P. (2019). Subchronic study of a white kidney bean (*Phaseolus vulgaris*) extract with α-amylase inhibitory activity. *BioMed Research International*, 2019. doi:10.1155/2019/9272345.

Ramírez-Jiménez, A. K., Gaytán-Martínez, M., Morales-Sánchez, E., & Loarca-Piña, G. (2018). Functional properties and sensory value of snack bars added with common bean flour as a source of bioactive compounds. *LWT*, *89*, 674-680. doi:10.1016/j.lwt.2017.11.043.

Rebollo-Hernanz, M., Martín-Cabrejas, M. A., & Aguilera, Y. (2018). Thermal processing of legumes. In Legumes (pp. 215-234).doi:10.1039/9781788015721-00215.

Rossi, G. B., Seraglio, S. K. T., Honaiser, T. C., Toaldo, I. M., Costa, A. C. D. O., Faria, J. C. D., & Arisi, A. C. M. (2022). Protein profile and antioxidant capacity of processed seeds from two common bean (*Phaseolus vulgaris* L.) cultivars. *International Journal of Food Science & Technology*, *57* (7), 3934-3943. doi:10.1111/ijfs.15537.

Roy, M., Imran, M. Z. H., Alam, M., & Rahman, M. (2021). Effect of boiling and roasting on physicochemical and antioxidant properties of dark red kidney bean (*Phaseolus vulgaris*). *Food Research*, *5* (3), 438-445. doi:10.26656/fr.2017.5(3).673.

Roy, M., Sarker, A., Azad, M. A. K., Shaheb, M. R., & Hoque, M. M. (2020). Evaluation of antioxidant and antimicrobial properties of dark red kidney bean (*Phaseolus vulgaris*) protein hydrolysates. *Journal of Food Measurement and Characterization*, *14*, 303-313. doi:10.1007/s11694-019-00292-4.

Sánchez-Chino, X. M., Corzo-Rios, L. J., Cid-Gallegos, M. S., & Jiménez-Martínez, C (2022). Compuestos antioxidantes provenientes de leguminosas como agentes quimiopreventivos en cáncer de colon. *Tequio, 5* (15), 111–124.

Stefanucci, A., Scioli, G., Marinaccio, L., Zengin, G., Locatelli, M., Tartaglia, A., Della Valle, A., Cichelli, A., Novellino, E., Pieretti, S., & Mollica, A. (2022). A comparative study on phytochemical fingerprint of two diverse *Phaseolus vulgaris* var. Tondino del Tavo and Cannellino Bio Extracts. *Antioxidants*, *11* (8), 1474. doi:10.3390/antiox11081474.

Teixeira-Guedes, C. I., Oppolzer, D., Barros, A. I., & Pereira-Wilson, C. (2019). Impact of cooking method on phenolic composition and antioxidant potential of four varieties of *Phaseolus vulgaris* L. and *Glycine max* L. *LWT, 103*, 238–246. doi:10.1016/j.lwt.2019.01.010.

Teodoro, A. J. (2019). Bioactive compounds of food: their role in the prevention and treatment of diseases. *Oxidative Medicine and Cellular Longevity, 2019*. doi:10.1155/2019/3765986.

Uebersax, M. A., Siddiq, M., Cramer, J., & Bales, S. (2022). Harvesting, postharvest handling, distribution, and marketing of dry beans. *Dry Beans and Pulses: Production, Processing, and Nutrition*, 81-104. doi:10.1002/9781119776802.ch4.

Wainaina, I., Wafula, E., Sila, D., Kyomugasho, C., Grauwet, T., Van Loey, A., & Hendrickx, M. (2021). Thermal treatment of common beans (*Phaseolus vulgaris* L.): Factors determining cooking time and its consequences for sensory and nutritional quality. *Comprehensive Reviews in Food Science and Food Safety, 20* (4), 3690-3718. doi:10.1111/1541-4337.12770.

Yang, Q. Q., Gan, R. Y., Ge, Y. Y., Zhang, D., & Corke, H. (2018). Polyphenols in common beans (*Phaseolus vulgaris* L.): Chemistry, analysis, and factors affecting composition. *Comprehensive Reviews in Food Science and Food Safety, 17* (6), 1518-1539. doi:10.1111/1541-4337.12391.

Yang, W., Zheng, Y., Sun, W., Chen, S., Liu, D., Zhang, H., & Ye, X. (2020). Effect of extrusion processing on the microstructure and *in vitro* digestibility of broken rice. *LWT, 119*, 108835.doi: /10.1016/j.lwt.2019.108835.

Zheng, Y., Liu, S., Xie, J., Chen, Y., Dong, R., Zhang, X., Liu, S., Xie, J., Hu, X., & Yu, Q. (2020). Antioxidant, α-amylase and α-glucosidase inhibitory activities of bound polyphenols extracted from mung bean skin dietary fiber. *LWT, 132*, 109943. doi:10.1016/j.lwt.2020.109943.

Zhou, R., Cai, W., & Xu, B. (2017). Phytochemical profiles of black and yellow soybeans as affected by roasting. *International Journal of Food Properties, 20* (12), 3179-3190. doi:10.1080/10942912.2017.1280678.

## Chapter 9

# The Role of Phenolic Compounds in the Antioxidant Activity of *Cicer arietinum*

**A. Linares-Castañeda[1]**
**L. J. Corzo-Ríos[2]**
**M. S. Cid-Gallegos[2]**
**Y. M. Gómez y Gómez[2]**
**X. M. Sánchez-Chino[3]**
**and C. Jiménez-Martínez[1,*]**
[1]Escuela Nacional de Ciencias Biológicas, Instituto Politécnico Nacional, Mexico City, Mexico
[2]Unidad Profesional Interdisciplinaria de Biotecnología, Instituto Politécnico Nacional, Mexico City, Mexico
[3]CONAHCyT, Departamento de Salud, El Colegio de la Frontera Sur-Villahermosa, Villahermosa, Tabasco, 86280, México

## Abstract

Legumes are an important source of macro and micro molecules in the human diet worldwide. Chickpea (*Cicer arietinum* L.) is one of the most cultivated legumes in its two types of varieties: Kabuli and Desi. These varieties differ in color, size, and chemical composition; however, both are protein sources (18 - 29%), starch (50 - 60%), lipids (4 - 7%), and bioactive compounds, such as phenolic compounds (0.72 - 1.81 mg/g seed). Most of the phenolic compounds are concentrated in the seed coat. Their concentration may vary depending on the type of variety and the extraction conditions, such as the type of solvent used, the extraction

* Corresponding Author's Email: crisjm_99@yahoo.com

In: Legumes: Beyond Their Nutritional Value
Editors: C. Jiménez-Martínez, M. F. Juárez-Chairez and M. S. Cid-Gallegos
ISBN: 979-8-89530-440-2

time, and temperature. Recent studies have documented the protective role of phenolic compounds, particularly their antioxidant properties, in mitigating damage caused by reactive oxygen species (ROS), including free radicals. These compounds have been shown to correlate with a reduced risk of cardiovascular diseases, diabetes, and certain types of cancer.
This is achieved by the expression and activity of antioxidant enzymes and by modifying the signal transduction pathways driven by the phenolic compounds in this legume. Therefore, studying and disseminating the benefits of legume consumption, particularly chickpeas, could positively impact human health.

**Keywords:** chickpea, phenolic compounds, antioxidant, human health

## 1. Introduction

Legumes belong to the *Leguminosae* (or *Fabaceae*) family and are considered the second most important family of plant crops after the grass family (*Poaceae*) (Bessada et al., 2019; Carbas et al., 2021). Legume seeds are dicotyledons and represent a relevant source of protein, dietary fiber, carbohydrates, vitamins, minerals, and phytochemicals (Bessada et al., 2019; Choi et al., 2019; Keskin et al., 2022). The differences between the nutritional composition and the content of phytochemicals among legumes are influenced by their genetics, variety, geographical location, and climatic conditions during their development; therefore, they have become integral foods in the human diet worldwide, they are cultivated in at least 57 countries and adapt to very diverse environmental conditions. In addition to their nutritional importance, their consumption helps to prevent cardiovascular disease, diabetes, and cancer (Carbas et al., 2021; Merga & Haji, 2019; Sharma, 2021).

Chickpea (*Cicer arietinum* L.) is one of the most economically relevant legumes, ranking fifth worldwide. India contributes 75% of its production, followed by Turkey, Pakistan, and Mexico; the latter ranks second as an exporter and third in production (Apáez Barrios et al., 2020). Nutritionally it contains 18-29% protein (rich in amino acids such as leucine, isoleucine, and arginine), 4-7% lipids, 50-60% starch, and 2-4% ash (Acquah et al., 2021; Boukid, 2021; Cid-Gallegos, Sánchez-Chino, Juárez Chairez, et al., 2020; Roorkiwal et al., 2020). The two chickpea varieties are the Kabuli and Desi types, which present differences in their chemical composition, color, size/weight, energy value, geographic area of cultivation, and flower color

(Boukid, 2021; Dida Bulbula & Urga, 2018; Juárez-Chairez et al., 2020; Kaur et al., 2019; Merga & Haji, 2019).

Regarding its nutritional chemical composition, there may be differences in moisture content, lipids, starch, ash, minerals, and vitamins between both varieties. Similarly, Kaur et al. (2019) have reported that the crude protein content was higher in the Kabuli-type chickpea (24.51%) and lower in the Desi type (20.29%); regarding the starch content, it was higher for the Kabuli-type (36.21-36.22 g/100 g) than for the Desi type (27.15-31.83 g/100 g); however, the lipid content and some minerals like zinc, iron, and magnesium, did not show an effect important to its nutritional composition. Additionally, it contains bioactive compounds (secondary metabolites) such as phenolic compounds, enzyme inhibitors, oligosaccharides, phytates, saponins, and lectins, among others. Phenolic compounds have shown advantages in preventing degenerative diseases attributed mainly to their antioxidant capacity (Cid-Gallegos et al., 2020; Perez-Perez et al., 2021; Yegrem, 2021).

## 2. Synthesis of Phenolic Compounds in Chickpea

Phenolic compounds (or polyphenols) are a group of secondary metabolites present in plants that have a wide range of biofunctional properties. Structurally, they have one or more hydroxyl groups attached directly to an aromatic ring, ranging from simple molecules to highly complex polymers. The synthesis of these compounds in legumes like chickpeas occurs in the testa and leaves through the shikimate pathway and metabolism of phenylpropanoids (Conti et al., 2021; Singh et al., 2017; Wojtunik-Kulesza et al., 2020).

The shikimate pathway aims to produce chorismate, the last common precursor in the aromatic amino acids L-phenylalanine, L-tyrosine, and L-tryptophan; however, polyphenols can only be derived from L-phenylalanine and L -tyrosine.

This pathway consists of seven enzymatic steps for the production of chorismate (Figure 1): 1) The pathway uses two molecules derived from primary metabolism, i.e., erythrose-4-phosphate (E4P) from the non-oxidative branch of the pathway of the pentose phosphate and phosphoenolpyruvate (PEP), produced by the glycolytic pathway, which is condensed to generate 3-deoxy-D-arabino-heptulosonic acid 7-phosphate (DAHP) by the action of 3-deoxy-D-arabino-heptulosonic acid 7-phosphate synthase (DAHPS), which requires $Mn^{2+}$ and reduced thioredoxin.

2) DAHP is cyclized to form 3-dehydroquinate (DHQ) by the action of 3-dehydrogenase synthase (DHQS). This reaction involves oxidation, phosphate removal, reduction, ring opening, and internal aldol condensation steps without forming byproducts.

3) and 4) DHQ undergoes two reactions that are combined and catalyzed by the bifunctional enzyme (unique in the shikimate pathway) 3-dehydroquinate dehydratase/shikimate dehydrogenase (DHQD/SDH) to form shikimate. DHQD (or dehydroquinase) initially catalyzes the dehydration of DHQ to 3-dehydroshikimate (DHS), which is subsequently reduced to shikimate by SDH with the use of nicotinamide adenine dinucleotide phosphate hydrogen (NADPH). A branch point that leads to the biosynthesis of quinate and other benzoates, such as protocatechuic and gallic acids, is located.

5) Shikimate originates shikimate 3-phosphate (S3P) by the action of shikimate kinase (SK), requiring divalent cations such as $Mg^{2+}$ and $Mn^{2+}$, transferring a phosphate group from ATP to the C3 hydroxyl group of shikimate.

6) Through the action of 5-enolpyruvylshikimate 3-phosphate synthase (EPSPS) and other PEP molecules are condensed to produce a 5-enolpyruvylshikimate 3-phosphate (EPSP) molecule, and finally,

7) Chorismate synthase (CS) dephosphorylates EPSP to produce chorismate (Gomes et al., 2022; Marchiosi et al., 2020).

For the biosynthesis of phenolic compounds, the first step consists of producing prephenate from chorismate by the action of chorismate mutase (CM). To produce L-phenylalanine, prephenate can be converted to phenylpyruvate or arogenate by prephenate dehydratase (PDT) or prephenate aminotransferase (PAT), respectively. In addition, prephenate can also be converted to p-hydroxyphenylpyruvate by prephenate dehydrogenase (PDH). Subsequently, L-tyrosine can be produced from p-hydroxyphenylpyruvate and arogenate by using aromatic amino-acid aminotransferase (AAT) and arogenate dehydrogenase (ADH), respectively. L-tyrosine and L-phenylalanine are converted to hydroxycinnamic acids via the phenylpropanoid pathway. When L-phenylalanine is used as a substrate, the first enzyme in this biosynthetic pathway is phenylalanine ammonia-lyase (PAL). This enzyme performs the deamination of L-phenylalanine to produce cinnamic acid. Then, cinnamic acid is hydroxylated by the enzyme cinnamate-4-hydroxylase (C4H) by adding a hydroxyl group at position 4 to produce p-coumaric acid. Moreover, simultaneously, p-coumaric acid could also be obtained in a single enzymatic step from L- tyrosine, which uses tyrosine ammonia lyase (TAL) to perform the deamination of L-tyrosine. Later, p-coumaric acid could be used to produce other hydroxycinnamic acids. 4-coumarate 3-hydroxylase (C3H) converts p-coumaric acid to caffeic acid (de la Rosa et al., 2019; Gomes et al., 2022).

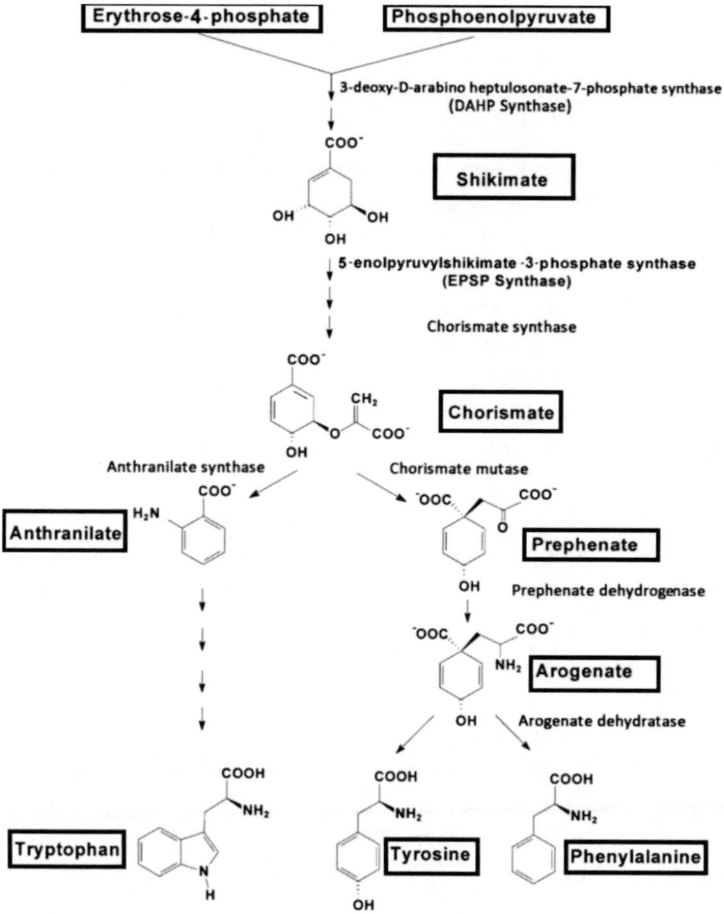


**Figure 1**. Shikimate pathway (Hanik et al., 2020).

Furthermore, caffeic acid could be converted to ferulic acid by caffeic acid 3-O-methyltransferase (COMT). These hydroxycinnamic acids could be activated to the corresponding coenzyme A (CoA) esters by 4-coumarate-CoA ligase (4CL). In addition, p-coumaroyl-CoA could also be converted to caffeoyl-CoA by p-coumaroyl 5-O-shikimate 3'-hydroxylase (CS3'H) or by p-coumaroyl shikimate transferase (CST). Subsequently, caffeoyl-CoA could be converted to feruloyl-CoA by caffeoyl-CoA 3-O methyltransferase (CCoAOMT). After the biosynthesis of CoA esters, the pathway is diverted toward forming different polyphenols. The enzymes involved in these steps differ depending on the polyphenol produced (de la Rosa et al., 2019; Gomes et al., 2022).

## 3. Phenolic Compounds in Chickpea

Highly pigmented legumes have higher phenolic compound values than less pigmented ones (Serventi & Dsouza, 2020). Therefore, it is expected that the darker the chickpea's color, the higher its content of phenolic compounds (thus, greater antioxidant properties) (Rachwa-Rosiak et al., 2015). The total content of phenolic compounds in chickpeas has been documented to range from 0.72 to 1.81 mg/g of seed. However, higher values have been reported (Table I); the variation in the relevance of these compounds is related to the type of chickpea variety (genetic factors), degree of maturation, seed size, environmental factors, and conditions during extraction (Rachwa-Rosiak et al., 2015; Summo et al., 2019).

Phenolic compounds are divided into five groups that, in turn, are made up of subgroups: flavonoids (flavonoids, flavonols, flavonoids, flavanols, isoflavones, anthocyanidins), phenolic acids (hydroxybenzoic acids and hydroxycinnamic acids), stilbenes (stilbene derivatives), tannins (hydrolyzable tannins and non-hydrolyzable tannins), and lignans (Câmara et al., 2020).

Hydroxybenzoic acid (HBA) and flavonoids, such as taxifolin and biochanin A, are the main phenolic compounds present in chickpeas (de Camargo et al., 2022; Wang et al., 2021). Depending on the variety of chickpeas, the concentration of the different phenolic compounds may vary; for example, the Kabuli type has a higher flavonoid content than the Desi type. On the other hand, phenolic acids and tannins have also been reported as important sources of phenolic compounds in both varieties, with higher content for the Desi type (Table 1). Stilbenes have been quantified regarding resveratrol in powdered juice of chickpea microgreens (Kumar et al., 2023) and sprouts (Ghumman et al., 2017).

**Table 1.** Content of phenolic compounds in chickpea

| Phenolic compound | Variety | Concentration | Reference |
|---|---|---|---|
| Flavonoids | Desi | 440-920[1] | Heiras-Palazuelos et al., 2013 |
| | | 1050.1[2] | Kose et al., 2022 |
| | | 587.7[3] | Domínguez-Arispuro et al., 2018 |
| | Kabuli | 2040[1] | Palazuelos et al., 2013 |
| | NS | 720[1] | Xu et al., 2007 |
| | | 84.3[3] | Saleh et al., 2019 |
| Phenolic acids | Desi | 107.0[4] | Magalhães et al., 2017 |
| | | 30.20[4] | Domínguez-Arispuro et al., 2018 |
| | Kabuli | 52.8[4] | Magalhães et al., 2017 |
| | | 27.5[4] | Magalhães et al., 2017 |
| | | 32.0[4] | Magalhães et al., 2017 |
| Stilbenes | NS | 710 – 3370[4] | Kumar et al., 2023 |
| | NS | 3197[4] | Ghumman et al., 2017 |
| Tanins | Desi | 520-3140[1] | Heiras-Palazuelos et al., 2013 |
| | Kabuli | 250[1] | Heiras-Palazuelos et al., 2013 |
| | NS | 520 [5] | Xu et al., 2007 |
| Lignans | NS | 3.73[4] | Durazzo et al., 2013 |

NS: Not specified. [1] mg CAE/kg (d.b.), [2] mg EAG/kg (d.b.), [3] mg quercetin/kg (d.b.), [4] mg/kg (d.b.). [5] mg CE/Kg.

## 4. Extraction and Quantification of Phenolic Compounds in Chickpea

Chickpea phenolic compounds have primarily been extracted using the maceration technique, which involves various solvents and solid-liquid extraction conditions (Table II). The efficiency of this extraction process is influenced by several factors, including particle size, solvent type, solute-to-solvent ratio, stirring speed, mass transfer efficiency, and extraction temperature (İşçimen & Hayta, 2021; Singh et al., 2017).

To extract phenolic compounds from chickpeas, methanol, ethanol, and acetone at different concentrations have been used; however, when absolute ethanol was used, it was impossible to extract the phenolic compounds (Table 2). In addition, sometimes, the extraction is carried out by modifying the pH of the solvent to provide stability to the phenolic compounds and increase their solubilization and diffusion through plant cells (Ahmed et al., 2020). It is also important to consider the extraction time since they have been used from a few minutes to hours; the consecutive number of extractions has also varied.

**Table 2.** Extraction parameters, total phenolic compounds (TPC), and antioxidant activity in chickpea

| Variety | Extraction parameters | TPC | Antioxidant activity | | References |
|---|---|---|---|---|---|
| Desi | 0.1 g/mL 80% EtOH/10 min, 50 rpm | 110.17 [1] | ORAC | 5456[b] | Domínguez-Arispuro et al., 2018 |
| | | | ABTS | 5871[b] | |
| | 0.1 g/mL EtOH al 80%/10 min. | 123 – 148[1] | ORAC | 5011 – 5756 [b] | Heiras-Palazuelos et al., 2013 |
| | 0.01 g/mL 50% Ace/12-16 h, AT, dark<br>0.01 g/mL 80% Ace/12-16 h, AT, dark<br>0.01 g/mL 70% MeOH/12-16 h, AT, dark<br>0.01 g/mL 70% EtOH/12-16 h, TA, dark | ~500[2]<br>~400[2]<br>~200[2]<br>~200[2] | FRAP | ~2000 [b]<br>~2000[b]<br>~900 [b]<br>~1000 [b] | Segev et al., 2010 |
| | 0.1 g/mL 50% MeOH / 1 h, AT | 20.5 [1] | | ND | Haileslassie et al., 2019 |
| | 0.1 g/mL MeOH/24 h, AT | 186.0 [1] | DPPH | 44.5 [a] | Mesfin et al., 2021 |
| | 0.1 g/mL 80% EtOH/10 min | 57[1] | ORAC | ~4000 [b] | Heiras-Palazuelos et al., 2013) |
| Kabuli | 0.25 g/ml 80% acetone/5 min | 333.3-379.2[1] | TEAC | ~11 – 35[b] | Arevalo et al., 2020 |
| | 0.25 g/ml 80% acetone/5 min | 333.3-379.2[1] | ORAC | ~39 [b] | Arevalo et al., 2020 |
| | | | FRAP | ~7.5 – 8.0 [b] | |
| | 0.1 g/mL MeOH/24 h, AT | 179.8 [1] | DPPH | 37.3 [a] | Mesfin et al., 2021 |
| | 0.1 g/mL 50% MeOH/1 h, AT | 13.7 [1] | | ND | Haileslassie et al., 2019 |
| | 1 g extracted three times with 80% MeOH | 119.50 [1] | DPPH | 21.10 [a] | Sofi et al., 2020 |
| | Two extractions 1) pure MeOH and 2) 80% MeOH | 470.92 [1] | | ND | Godrich et al., 2023 |
| Not specified | Acetone/EtOH/water 40:40:20 v/v | 45.5 [1] | ABTS | 953.9 [b] | León-López et al., 2020 |
| | 0.1 g/mL 50% Ace/3 h, AT | 181 [1] | DPPH | 1.05 [b] | Xu & Chang, 2007 |
| | 0.1 g/mL 50% Ace/3 h, AT<br>0.1 g/mL 80% Ace/3 h, AT | 181 [1]<br>141 [1] | FRAP | 730 [d] | Xu & Chang, 2007 |
| | 0.1 g/mL 50%Ace/3 h, AT | 181 [1] | ORAC | 513 [b] | Xu & Chang, 2007 |
| | 0.1 g/mL 80% Ace/3 h, AT | 141 [1] | DPPH | 98 [b] | |
| | | 141 [1] | FRAP | 1.08 [d] | |
| | 0.1 g/mL acetone:water :AcOH 70:29.5:0.5 v/v/v/3 h, AT | 157[1] | | | |

| Variety | Extraction parameters | TPC | Antioxidant activity | | References |
|---|---|---|---|---|---|
| Not specified | 0.1 g/mL 80% Ace/3 h, AT | 141 [1] | ORAC | 1000 [b] | Xu & Chang, 2007 |
| | 0.1 g/mL Ace:water :AcOH 70:29.5:0.5 v/v/v/3 h, AT | 157[1] | DPPH | 94 [b] | |
| | | 157[1] | FRAP | 0.81 [d] | |
| | 0.1 g/mL 70% MeOH/3 h, AT | 168 [1] | | | |
| | 0.1 g/mL Ace:water :AcOH 70:29.5:0.5 v/v/v/3 h, AT | 157[1] | ORAC | 741 [b] | Xu & Chang, 2007 |
| | 0.1 g/mL 70% MeOH/3 h, AT | 168 [1] | DPPH | 236 [b] | |
| | | 168 [1] | FRAP | 0.91 [d] | |
| | | 154[1] | | | |
| | 0.1 g/mL 70% MeOH/3 h, AT | 168 [1] | ORAC | 2201 [b] | Xu & Chang, 2007 |
| | | 154[1] | DPPH | 206 [b] | |
| | | 154[1] | FRAP | 1.13 [d] | |
| | 0.1 g/mL absolute EtOH/3 h, AT | ND | | | |
| | 0.1 g/mL 70% EtOH/3 h, AT | 154[1] | ORAC | 3466 [b] | Xu & Chang, 2007 |
| | 0.1 g/mL absolute EtOH/3h, AT | ND | DPPH | 30 [b] | |
| | | ND | FRAP | 0.90 [d] | |
| | 0.1 g/mL acetone: water (50:50 v/v)/15 h, AT, dark | 98[1] | | | Xu et al., 2007 |
| | 0.1 g/mL absolute EtOH/3 h, AT | ND | ORAC | 3227 [b] | Xu & Chang, 2007 |
| | 0.1 g/mL acetone: water (50:50 v/v)/15 h, AT, dark | 98[1] | DPPH | 126 [b] | Xu et al., 2007 |
| | | 98[1] | FRAP | 0.80 [d] | |
| | 0.2 g/mL MeOH/45 min dark and sonication + two washes of the solid residue with methanol | 568 [1] | | | Saleh et al., 2019 |
| | 0.1 g/mL acetone: water (50:50 v/v)/15 h, AT, dark | 98[1] | ORAC | 926 [b] | Xu et al., 2007 |
| | 0.2 g/1 mL MeOH/45 min dark and sonication + two washes of the solid residue with methanol. | 568 [1] | DPPH | 86.70 [a] | Saleh et al., 2019 |
| | 0.05g/mL HCl/MeOH /water (1:80:10mL)/2 h, 25°C | 802 [1] | DPPH | 22.85 [a] | Jogihalli et al., 2017 |
| | 0.25 g/mL 70% MeOH/2 h, AT | 14.24[1] | FRAP | 5.47 [c] | Lafarga et al., 2019 |

**Table 2.** (Continued)

| Variety | Extraction parameters | TPC | Antioxidant activity | | References |
|---|---|---|---|---|---|
| Not specified | 0.25 g/mL 70% MeOH/2 h, AT | 14.24[1] | DPPH | 3.51 [c] | Lafarga et al., 2019 |
| | 0.2 g mL 70% EtOH/3 h, 65°C | ND | | | Kim et al., 2022 |
| | 0.25 g/ mL70% MeOH/2 h, AT | 14.24[1] | | ND | Lafarga et al., 2019 |
| | 0.2 g/mL 70% EtOH/3 h, 65°C | ND | | | Kim et al., 2022 |

ND: Not determined. AT = room temperature. [1]mg gallic acid equivalents/100 g (d.b.), [2]mg (+) catechin equivalents/100 g, [a]% inhibition, [b]µmol Trolox equivalents (TE)/100 g (d. b.), [c]mg ascorbic acid equivalents/100 g (d.b.) [d]µmol of $Fe^{2+}$ equivalents/100 g. AcOH Acetic acid, Ace= acetone.

On the other hand, light and temperature are also critical factors to consider during extraction because some of the phenolic compounds can degrade due to their sensitivity to light or their thermolabile nature (Osorio-Tobón, 2020); therefore, the extraction temperatures of phenolic compounds in chickpea range between 25 – 65°C, sometimes protected from light.

Fast, safe, and environmentally sustainable methods have been developed because these extraction techniques require large amounts of solvents and long extraction times. Among these methods are ultrasound-assisted, microwave-assisted, supercritical fluid extraction, and extraction with pressurized liquids, all leading to reduced use of organic solvents and extraction times (Shang et al., 2019). Microwave-assisted extraction with an ethanol/ammonium sulfate two-phase aqueous system has been implemented to extract phenolic compounds in chickpeas, using less solvent, shorter extraction time, and greater yield and antioxidant properties (İşçimen & Hayta, 2021). For this reason, no process for extracting phenolic compounds from chickpeas is considered universal. However, it is important always to establish the same conditions and extraction parameters to reduce possible variations during their determination.

On the other hand, techniques such as the Folin-Ciocalteu assay are used to quantify total phenolic compounds. This technique is based on the reaction of a mixture of phosphotungstic ($H_3PW_{12}O_{40}$) and phosphomolybdic ($H_3PMo_{12}O_{40}$) acids, which, in the presence of phenolic compounds, are reduced to oxides of tungsten blue ($W_8O_{23}$) and molybdenum blue ($Mo_8O_{23}$); this reduction can be detected with a spectrophotometer in the range from 500 to 750 nm (Calabria et al., 2020; Carrasco-Pancorbo et al., 2005), the reading is compared using a curve elaborated with a phenolic compound reference that is generally gallic acid (Xu & Chang, 2007). This technique is the most recognized for quantifying phenolic compounds, as it is reproducible and straightforward. Another technique used is liquid chromatography, which quantifies phenolic compounds and separates and identifies them (López-Fernández et al., 2020).

## 5. Antioxidant Activity of Chickpea Phenolic Compounds

Antioxidants delay auto-oxidation by inhibiting the formation of free radicals (RL) or interrupting their propagation. RLs, including reactive oxygen and nitrogen species (ROS and RNS), are atoms, molecules, or ions with one or more unpaired electrons in the outermost orbital, making them highly unstable (due to lack of octet configuration) and active against chemical reactions with other molecules (oxygen, nitrogen, and sulfur derivatives). To stabilize these molecules, antioxidants can intervene through one or more mechanisms: a) purifying species that initiate peroxidation, b) quench·$O_2^-$ preventing the formation of peroxides, c) breaking the auto-oxidation chain reaction, d) reducing the localized $O_2$ concentrations, among others (Brewer, 2011; Gulcin, 2020). These reactions are crucial in human metabolism because EROS, ERNS, and RL can initiate some diseases affecting the quality of life (de Camargo et al., 2019; García-Caparrós et al., 2021; Munteanu & Apetrei, 2021).

The human body relies on antioxidant systems to scavenge these free radicals, and under normal conditions, the speed and extent of formation are balanced by the rate of their scavenging. However, losing the balance between pro-oxidant and antioxidant molecules generates oxidative stress. This uncontrolled increase in EROS, ERNS, and RL disrupts the conformation of nucleic acids through oxidation of deoxyribose, removal/deletion of nucleotides, base modifications, and cross-linking of proteins with DNA. It causes lipid peroxidation by breaking longer chains, increasing the fluidity and permeability of the membranes, and oxidizing proteins, generating modifications such as cleavage of peptide chains, cross-linking of proteins, and changing the electrical charge (de Camargo et al., 2019; García-Caparrós et al., 2021; Munteanu & Apetrei, 2021).

Due to the molecular structure of phenolic compounds (one or more aromatic rings containing one or more hydroxyl groups), they can neutralize free radicals by forming resonance-stabilized phenoxyl radicals (Koley et al., 2019). Therefore, in legumes, phenolic compounds (particularly isoflavones) have been reported as responsible for the antioxidant properties, as they can act at the cell membrane level and break the chain reaction of RLs on the cell surface. They can cross the membrane and react with intracellular ROS, in addition to being able to modulate the activity of antioxidant enzymes such as catalase, superoxide dismutase, glutathione peroxidase, glutathione reductase, and peroxiredoxins (de Camargo et al., 2019; Domínguez-Arispuro et al., 2021; Koley et al., 2019; Wang et al., 2021).

The chemical structure is essential for the antioxidant activity of flavonoids because a certain pattern of hydroxylation, particularly in the B ring, and/or double bond between C2=C3 in conjugation with a C4 carbonyl group, methoxyl groups, and O3-H increase antioxidant activity. Flavonoids' ability to eliminate RLs is based on the B ring with an ortho-dihydroxy group (catechol) (Vo et al., 2019; Wojtunik-Kulesza et al., 2020).

The values of antioxidant activity in chickpeas will depend on the type of test used, which can be through 2,2-diphenyl-1-picrylhydrazyl (DPPH), 2,2'-azino-bis (3-ethylbenzothiazoline-6-sulfonic acid (ABTS), oxygen radical absorbance capacity (ORAC), and ferric ion reducing activity (FRAP); the type of chickpea (Desi/Kabuli) and the extraction conditions of the compounds with antioxidant activity. For example, DPPH and ORAC tests have shown that the Desi-type variety has higher antioxidant activity than the Kabuli-type variety. Likewise, it has been shown that the type of solvent during the extraction can modify the antioxidant activity; in the case of the DPPH test, lower antioxidant activity was obtained in chickpeas using 80% acetone, compared to the extracts with 70% ethanol and methanol (Table II). However, a correlation must be made between the different tests (DPPH, ABTS, FRAP, ORAC) to determine the optimal conditions for the best extraction of phenolic compounds and, as a result, the greatest antioxidant activity.

In addition, it is important to consider that the seeds can present antioxidant properties and appear in chickpea sprouts, pods, peels, and roots (Wang et al., 2021). It has been shown that chickpea husks have higher antioxidant activity compared to shelled chickpea seeds (Mahbub et al., 2021). Other studies have shown that sprouting can significantly increase antioxidant activity and positively correlate with increasing sprouting time. For example, Sofi et al. (2020) showed an increase in the concentration of phenolic compounds in the sprouts compared to the raw seeds; they reported 119.50 and 254.25 mg EAG/100 g at 0 and 48 h of germination, respectively.

## *6. In Vitro* and *In Vivo* Studies of Chickpea Phenolic Compounds

As previously mentioned, phytochemicals such as phenolic compounds can regulate the redox balance by targeting the Nrf2 pathway and prevent cardiovascular diseases, some types of cancer, neurodegenerative diseases, diabetes, obesity, aging, and other chronic inflammatory disorders (He et al., 2023).

Through *in vivo* studies, it has been shown that the Desi-type chickpea peel improves the antioxidant status in rats due to its content of phenolic compounds, offering protection against oxidation of blood lipids, which again translates as a risk factor for developing cardiovascular disease. It was also shown that chickpea peel increases the concentration of short-chain fatty acids (butyrate), helping the proliferation of intestinal epithelial cells and modulating cecal fermentation, resulting in positive alteration of intestinal flora (Myint et al., 2017).

On the other hand, *in vitro* studies have shown that polyphenolic extracts from the chickpea peel decrease oxidative stress in cells because catalase and glutathione peroxidase activity was increased; in addition, they reduce the production of inflammatory markers such as nitric oxide and interleukin-6 (which is a cytokine expressed at low levels, however, during infections, trauma, aging or under stress conditions it is over-expressed) (Mahbub et al., 2021; Wang et al., 2021).

Similarly, flavonoids extracted from raw Kabuli chickpeas attenuate oxidative stress in rats with type 2 diabetes mellitus (Fu et al., 2022). Biochanin A, present in chickpeas, is the flavonoid that has received the most attention due to its therapeutic value. It has been shown to reduce oxidative stress in diabetic rats (de Camargo et al., 2019). In addition, other biofunctional properties may be present in chickpea extracts, such as the anti-inflammatory activity demonstrated in murine models (Sandeep et al., 2012).

Studies on cell lines using isoflavones from chickpea sprouts revealed a great potential in cellular antioxidant activity and helping in chemoprevention or chemotherapy against breast cancer (Chen et al., 2015; Domínguez-Arispuro et al., 2021). Likewise, in models of diabetic rats, the methanolic extracts of chickpea sprouts (with 352.98 μg/mg dry sample) induced a decrease in cholesterol, triglycerides, and low-density lipoprotein cholesterol (LDL-C). The phenolic extracts increased high-density lipoprotein cholesterol (HDL-C) levels, further demonstrating their protective effects against cardiovascular diseases (Farag et al., 2023).

Other processes, such as cooking, have enhanced the effect of inhibition of cancer cell proliferation and the chemopreventive effect on colon cancer; however, it is important to consider the chickpea variety since the Desi variety has shown greater potential for the inhibition of cell proliferation than the Kabuli variety (Cid-Gallegos et al., 2020; Milán-Noris et al., 2023; Sánchez-Chino et al., 2017).

It is important to consider that during the processing of these foods (for example, soaking, grinding, cooking, roasting, etc.), the antioxidant properties may decrease due to the loss of phenolic compounds; for example, the cooking process reduces the content of phenolic compounds due to their thermal sensitivity or their leaching into the cooking water. For this reason, some of the research has focused on developing and sensory acceptance of these new products and conserving this biofunctionality (Giusti et al., 2019; Kaur & Prasad, 2021).

The boom in research for preventing these diseases through consuming foods rich in phenolic compounds continues constantly. Legume seeds, including chickpeas, are a promising option for developing functional foods to promote their consumption and human health, as the per capita consumption of legumes is low despite the multiple benefits of consuming this type of food (Keskin et al., 2022).

Due to the *in vivo* and *in vitro* evidence of the benefits of chickpea consumption in the prevention of some chronic degenerative diseases, various studies have developed chickpea-based foods. Some of these investigations include the elaboration of extruded chickpea/sorghum, chickpea/rice feeds; gluten-free crackers made from chickpea/rice flour, chickpea/chestnut, sprouted chickpea; chickpea/acorn flour gluten-free bread; pasta with chickpea flour/wheat semolina and chickpea flour tortillas (Altaf et al., 2022; Bekele et al., 2020; Bon-Padilla et al., 2022; El-Sohaimy et al., 2020; Garcia-Valle et al., 2021; Gkountenoudi-Eskitzi et al., 2023; Hamdani et al., 2020; Singh Sibian & Singh Riar, 2021; Torra et al., 2021). Synergies have also been developed between chickpea-based foods and other important sources of phenolic compounds, such as a fried snack based on chickpea flour with onion peel extract (Siddiqui et al., 2022).

For chickpea/sorghum snacks, it was found that a higher barrel temperature and a higher proportion of sorghum in the formulation increased the inhibition of DPPH radicals and the total content of phenols, preserving the highest antioxidant capacity and phenols. In total, it is suggested that the snacks be made with a 50:50 chickpea: sorghum mixture extruded at 160°C and with a humidity of 16% (Bekele et al., 2021). Likewise, the dietary fiber extracted from the chickpea peel was included in the formulation of white bread, showing antioxidant activity after baking (Niño-Medina et al., 2019). On the other hand, in addition to preserving biofunctional properties after processing, techno-functional properties can simultaneously occur in developing new foods, such as sprouted chickpea biscuits, which present greater protein and carbohydrate digestibility (Singh Sibian & Singh Riar, 2021).

For this reason, it is important to know if, after preparing this type of food, the phenolic compounds are still preserved and help prevent chronic degenerative diseases.

## Conclusion

In this chapter, phenolic compounds in chickpea (*Cicer arietinum* L.) have been documented, along with some extraction and quantification methods and other compounds responsible for the antioxidant activity to protect against cell damage caused by free radicals (oxidative stress). In general terms, during the determination of phenolic compounds, the type of chickpea (Desi/Kabuli) and the conditions during the extraction can influence their quantification. Therefore, comparing the different studies to establish the best conditions for phenolic compound extraction and, consequently, greater antioxidant activity in chickpeas is important. On the other hand, chickpeas have gained attention due to the ample evidence in reducing chronic-degenerative diseases. Therefore, its inclusion in the diet is promising to improve health; however, despite these advantages, the consumption of this type of legume is low, so there are still challenges to its incorporation into the formulation of foods to promote its consumption and preserve the antioxidant properties after processing.

## References

Acquah, C., Ohemeng-Boahen, G., Power, K. A., & Tosh, S. M. (2021). The effect of processing on bioactive compounds and nutritional qualities of pulses in meeting the sustainable development goal 2. *Frontiers in Sustainable Food Systems*, *5*. doi:10.3389/fsufs.2021.681662.

Ahmed, M. I., Xu, X., Sulieman, A. A., Mahdi, A. A., & Na, Y. (2020). Effect of extraction conditions on phenolic compounds and antioxidant properties of koreeb (*Dactyloctenium aegyptium*) seeds flour. *Journal of Food Measurement and Characterization*, *14*(2), 799–808. doi:10.1007/s11694-019-00328-9.

Altaf, U., Hussain, S. Z., Naseer, B., Amin, T., & Bashir, O. (2022). Nutritional and bioactive components of rice-chickpea-based snacks as affected by severe and mild extrusion cooking. *Journal of the Science of Food and Agriculture*, *102*(15), 7126–7135. doi:10.1002/jsfa.12075.

Apáez Barrios, M., Escalante Estrada, J. A. S., Apáez Barrios, P., & Alvarez Hernandez, J. C. (2020). Producción, crecimiento y calidad nutrimental del garbanzo en función del nitrógeno y fósforo. *Revista Mexicana de Ciencias Agrícolas*, *11*(6), 1273–1284. doi:10.29312/remexca.v11i6.2226.

Arevalo, I., Guzmán-Maldonado, S. H., Sanchez, S. M. M., & Acosta-Gallegos, J. A. (2020). Steaming and toasting reduce the nutrimental quality, total phenols and antioxidant capacity of fresh kabuli Chickpea (*Cicer arietinum* L.). *Plant Foods for Human Nutrition*, *75*(4), 628–634. doi:10.1007/s11130-020-00857-5.

Bekele, E. K., Nosworthy, M. G., Henry, C. J., Shand, P. J., & Tyler, R. T. (2020). Oxidative stability of direct-expanded chickpea–sorghum snacks. *Food Science & Nutrition, 8*(8), 4340–4351. doi:10.1002/fsn3.1731.

Bekele, E. K., Nosworthy, M. G., Tyler, R. T., & Henry, C. J. (2021). Antioxidant capacity and total phenolics content of direct-expanded chickpea–sorghum snacks. *Journal of Food Processing and Preservation, 45*(5), e15439. doi:10.1111/jfpp.15439.

Bessada, S. M. F., Barreira, J. C. M., & Oliveira, M. B. P. P. (2019). Pulses and food security: Dietary protein, digestibility, bioactive and functional properties. *Trends in Food Science & Technology, 93*, 53–68. doi:10.1016/j.tifs.2019.08.022.

Bon-Padilla, B. K., Reyes-Moreno, C., Milán-Carrillo, J., Reynoso-Camacho, R., Gómez-Aldapa, C. A., Gómez-Favela, M. A., & Gutiérrez-Dorado, R. (2022). Tortillas made from nixtamalized maize and extruded chickpea flours: A product with improved *in vitro* nutritional and antihypertensive properties. *Cereal Chemistry, 99*(5), 1154–1165. doi:10.1002/cche.10577.

Boukid, F. (2021). Chickpea (*Cicer arietinum* L.) protein as a prospective plant-based ingredient: a review. *International Journal of Food Science & Technology, 56*(11), 5435–5444. doi:10.1111/ijfs.15046.

Brewer, M. S. (2011). Natural antioxidants: Sources, compounds, mechanisms of action, and potential applications. *Comprehensive Reviews in Food Science and Food Safety, 10*(4), 221–247. doi:10.1111/j.1541-4337.2011.00156.x.

Calabria, D., Mirasoli, M., Guardigli, M., Simoni, P., Zangheri, M., Severi, P., Caliceti, C., & Roda, A. (2020). Paper-based smartphone chemosensor for reflectometric on-site total polyphenols quantification in olive oil. *Sensors and Actuators B: Chemical, 305*, 127522. doi:10.1016/j.snb.2019.127522.

Câmara, J. S., Albuquerque, B. R., Aguiar, J., Corrêa, R. C. G., Gonçalves, J. L., Granato, D., Pereira, J. A. M., Barros, L., & Ferreira, I. C. F. R. (2020). Food bioactive compounds and emerging techniques for their extraction: polyphenols as a case study. *Foods, 10*(1), 37. doi:10.3390/foods10010037.

Carbas, B., Machado, N., Pathania, S., Brites, C., Rosa, E. A. S., & Barros, A. I. (2021). Potential of legumes: nutritional value, bioactive properties, innovative food products, and application of eco-friendly tools for their assessment. *Food Reviews International*, 1–29. doi:10.1080/87559129.2021.1901292.

Carrasco-Pancorbo, A., Cerretani, L., Bendini, A., Segura-Carretero, A., Gallina-Toschi, T., & Fernández-Gutiérrez, A. (2005). Analytical determination of polyphenols in olive oils. *Journal of Separation Science, 28*(9–10), 837–858. doi:10.1002/jssc.200500032.

Chen, H., Ma, H.-R., Gao, Y.-H., Zhang, X., Habasi, M., Hu, R., & Aisa, H. A. (2015). Isoflavones extracted from chickpea *Cicer arietinum* L. sprouts induce mitochondria-dependent apoptosis in human breast cancer cells. *Phytotherapy Research, 29*(2), 210–219. doi:10.1002/ptr.5241.

Choi, W. C., Parr, T., & Lim, Y. S. (2019). The impact of four processing methods on trypsin-, chymotrypsin- and alpha-amylase inhibitors present in underutilized legumes. *Journal of Food Science and Technology, 56*(1), 281–289. doi:10.1007/s13197-018-3488-0.

Cid-Gallegos, M. S., Sánchez-Chino, X. M., Álvarez-González, I., Madrigal-Bujaidar, E., Vásquez-Garzón, V. R., Baltiérrez-Hoyos, R., Villa-Treviño, S., Dávila-Ortíz, G., & Jiménez-Martínez, C. (2020). Modification of *in vitro* and *in vivo* antioxidant activity by consumption of cooked chickpea in a colon cancer Model. *Nutrients*, *12*(9). doi:10.3390/nu12092572.

Cid-Gallegos, M. S., Sánchez-Chino, X. M., Juárez Chairez, M. F., Álvarez González, I., Madrigal-Bujaidar, E., & Jiménez-Martínez, C. (2020). Anticarcinogenic activity of phenolic compounds from sprouted legumes. *Food Reviews International*, 1–16. doi:10.1080/87559129.2020.1840581.

Conti, M. V., Guzzetti, L., Panzeri, D., De Giuseppe, R., Coccetti, P., Labra, M., & Cena, H. (2021). Bioactive compounds in legumes: Implications for sustainable nutrition and health in the elderly population. *Trends in Food Science & Technology*, *117*, 139–147. doi:10.1016/j.tifs.2021.02.072.

de Camargo, A. C., Alvarez, A. C., Arias-Santé, M. F., Oyarzún, J. E., Andia, M. E., Uribe, S., Pizarro, P. N., Bustos, S. M., Schwember, A. R., Shahidi, F., & Bridi, R. (2022). Soluble free, esterified and insoluble-bound phenolic antioxidants from chickpeas prevent cytotoxicity in human hepatoma HuH-7 Cells induced by peroxyl radicals. *Antioxidants*, *11*(6). doi:10.3390/antiox11061139.

de Camargo, A. C., Favero, B. T., Morzelle, M. C., Franchin, M., Alvarez-Parrilla, E., de la Rosa, L. A., Geraldi, M. V., Maróstica Júnior, M. R., Shahidi, F., & Schwember, A. R. (2019). Is chickpea a potential substitute for soybeans? phenolic bioactives and potential health benefits. *International Journal of Molecular Sciences*, *20*(11), 2644. doi:10.3390/ijms20112644.

de la Rosa, L. A., Moreno-Escamilla, J. O., Rodrigo-García, J., & Alvarez-Parrilla, E. (2019). Phenolic Compounds. In E. M. Yahia (Ed.), *Postharvest physiology and biochemistry of fruits and vegetables* Chapter 12 - (pp. 253–271). Woodhead Publishing. doi:10.1016/B978-0-12-813278-4.00012-9.

Dida Bulbula, D., & Urga, K. (2018). Study on the effect of traditional processing methods on nutritional composition and anti nutritional factors in chickpea (*Cicer arietinum*). *Cogent Food & Agriculture*, *4*(1), 1422370. doi:10.1080/23311932.2017.1422370.

Domínguez-Arispuro, D. M., Cuevas-Rodríguez, E. O., Milán-Carrillo, J., León-López, L., Gutiérrez-Dorado, R., & Reyes-Moreno, C. (2018). Optimal germination condition impacts on the antioxidant activity and phenolic acids profile in pigmented desi chickpea (*Cicer arietinum* L.) seeds. *Journal of Food Science and Technology*, *55*(2), 638–647. doi:10.1007/s13197-017-2973-1.

Domínguez-Arispuro, D-M, Adrián, C.-R., Cuauhtémoc, R.-M., Ada-Keila, M.-N., Jorge, M.-C., Erika, A.-S., & Edith-Oliva, C.-R. (2021). Isoflavones from black chickpea (*Cicer arietinum* L) sprouts with antioxidant and antiproliferative activity. *Saudi Journal of Biological Sciences*, *28*(1), 1141–1146. doi:10.1016/j.sjbs.2020.11.048.

Durazzo, A., Turfani, V., Azzini, E., Maiani, G., & Carcea, M. (2013). Phenols, lignans and antioxidant properties of legume and sweet chestnut flours. *Food Chemistry*, *140*(4), 666–671. doi:10.1016/j.foodchem.2012.09.062.

El-Sohaimy, S. A., Brennan, M., Darwish, A. M. G., & Brennan, C. (2020). Physicochemical, texture and sensorial evaluation of pasta enriched with chickpea flour and protein isolate. *Annals of Agricultural Sciences*, *65*(1), 28–34. doi:10.1016/j.aoas.2020.05.005.

Farag, M. A., Aboul Naser, A. F., Zayed, A., & Sharaf El-Dine, M. G. (2023). Comparative insights into four major legume sprouts efficacies for diabetes management and its complications: untargeted versus targeted NMR biochemometrics approach. *Metabolites*, *13*(1). doi:10.3390/metabo13010063.

Fu, Y., Li, Z., Xiao, S., Zhao, C., Zhou, K., & Cao, S. (2022). Ameliorative effects of chickpea flavonoids on redox imbalance and mitochondrial complex I dysfunction in type 2 diabetic rats. *Food & Functional*, *13*(17), 8967–8976. doi:10.1039/D2FO00753C.

García-Caparrós, P., De Filippis, L., Gul, A., Hasanuzzaman, M., Ozturk, M., Altay, V., & Lao, M. T. (2021). Oxidative stress and antioxidant metabolism under adverse environmental conditions: a Review. *The Botanical Review*, *87*(4), 421–466. doi:10.1007/s12229-020-09231-1.

Garcia-Valle, D. E., Bello-Pérez, L. A., Agama-Acevedo, E., & Alvarez-Ramirez, J. (2021). Structural characteristics and in vitro starch digestibility of pasta made with durum wheat semolina and chickpea flour. *LWT*, *145*, 111347. doi:10.1016/j.lwt.2021.111347.

Ghumman, A., Singh, N., & Kaur, A. (2017). Chemical, nutritional and phenolic composition of wheatgrass and pulse shoots. *International Journal of Food Science & Technology*, *52*(10), 2191–2200. doi:10.1111/ijfs.13498.

Giusti, F., Capuano, E., Sagratini, G., & Pellegrini, N. (2019). A comprehensive investigation of the behaviour of phenolic compounds in legumes during domestic cooking and in vitro digestion. *Food Chemistry*, *285*, 458–467. doi:10.1016/j.foodchem.2019.01.148.

Gkountenoudi-Eskitzi, I., Kotsiou, K., Irakli, M. N., Lazaridis, A., Biliaderis, C. G., & Lazaridou, A. (2023). *In vitro* and in *vivo* glycemic responses and antioxidant potency of acorn and chickpea fortified gluten-free breads. *Food Research International*, *166*, 112579. doi:10.1016/j.foodres.2023.112579.

Godrich, J., Rose, P., Muleya, M., & Gould, J. (2023). The effect of popping, soaking, boiling and roasting processes on antinutritional factors in chickpeas and red kidney beans. *International Journal of Food Science & Technology*, *58*(1), 279–289. doi:10.1111/ijfs.16190.

Gomes, D., Rainha, J., Rodrigues, L. R., & Rodrigues, J. L. (2022). Yeast Synthetic Biology Approaches for the Production of Valuable Polyphenolic Compounds. In *Synthetic Biology of Yeasts: Tools and Applications.* F. Darvishi Harzevili (Ed.),pp. 119–156. Springer International Publishing. doi:10.1007/978-3-030-89680-5_5.

Gulcin, İ. (2020). Antioxidants and antioxidant methods: an updated overview. *Archives of Toxicology*, *94*(3), 651–715. doi:10.1007/s00204-020-02689-3.

Haileslassie, H. A., Henry, C. J., & Tyler, R. T. (2019). Impact of pre-treatment (soaking or germination) on nutrient and anti-nutrient contents, cooking time and acceptability of cooked red dry bean (*Phaseolus vulgaris* L.) and chickpea (*Cicer arietinum* L.) grown in Ethiopia. *International Journal of Food Science & Technology*, *54*(8), 2540–2552. doi:10.1111/ijfs.14165.

Hamdani, A. M., Wani, I. A., & Bhat, N. A. (2020). Gluten free cookies from rice-chickpea composite flour using exudate gums from acacia, apricot and karaya. *Food Bioscience*, *35*, 100541. doi:10.1016/j.fbio.2020.100541.

Hanik, N., Best, M., Schueller, M. J., Tappero, R., & Ferrieri, R. A. (2020). Defense Priming in Nicotiana tabacum Accelerates and Amplifies 'New' C/N Fluxes in Key Amino Acid Biosynthetic Pathways. *Plants*, *9*(7). doi: 10.3390/plants9070851.

He, W.-J., Lv, C.-H., Chen, Z., Shi, M., Zeng, C.-X., Hou, D.-X., & Qin, S. (2023). The Regulatory Effect of Phytochemicals on Chronic Diseases by Targeting Nrf2-ARE Signaling Pathway. *Antioxidants*, *12*(2), 236. doi:10.3390/antiox12020236.

Heiras-Palazuelos, M. J., Ochoa-Lugo, M. I., Gutiérrez-Dorado, R., López-Valenzuela, J. A., Mora-Rochín, S., Milán-Carrillo, J., Garzón-Tiznado, J. A., & Reyes-Moreno, C. (2013). Technological properties, antioxidant activity and total phenolic and flavonoid content of pigmented chickpea (*Cicer arietinum* L.) cultivars. *International Journal of Food Sciences and Nutrition*, *64*(1), 69–76. doi:10.3109/09637486.2012.694854.

İşçimen, E. M., & Hayta, M. (2021). Microwave-assisted aqueous two-phase system based extraction of phenolics from pulses: Antioxidant properties, characterization and encapsulation. *Industrial Crops and Products*, *173*, 114144. doi:10.1016/j.indcrop.2021.114144.

Jogihalli, P., Singh, L., Kumar, K., & Sharanagat, V. S. (2017). Physico-functional and antioxidant properties of sand-roasted chickpea (*Cicer arietinum*). *Food Chemistry*, *237*, 1124–1132. doi:10.1016/j.foodchem.2017.06.069.

Juárez-Chairez, M. F., Cid-Gallegos, M. S., Meza-Márquez, O. G., & Jiménez-Martínez, C. (2020). Biological Activities of Chickpea in Human Health (*Cicer arietinum* L.). A Review. *Plant Foods for Human Nutrition*, *75*(2), 142–153. doi:10.1007/s11130-020-00814-2.

Kaur, K., Grewal, S. K., Gill, P. S., & Singh, S. (2019). Comparison of cultivated and wild chickpea genotypes for nutritional quality and antioxidant potential. *Journal of Food Science and Technology*, *56*(4), 1864–1876. doi:10.1007/s13197-019-03646-4.

Kaur, R., & Prasad, K. (2021). Technological, processing and nutritional aspects of chickpea (*Cicer arietinum*) - A review. *Trends in Food Science & Technology*, *109*, 448–463. doi:10.1016/j.tifs.2021.01.044.

Keskin, S. O., Ali, T. M., Ahmed, J., Shaikh, M., Siddiq, M., & Uebersax, M. A. (2022). Physico-chemical and functional properties of legume protein, starch, and dietary fiber—A review. *Legume Science*, *4*(1), e117. doi:10.1002/leg3.117.

Kim, S. M., Aung, T., & Kim, M. J. (2022). Optimization of germination conditions to enhance the antioxidant activity in chickpea (*Cicer arietinum* L.) using response surface methodology. *Korean Journal of Food Preservation*, *29*(4), 632–644. doi:10.11002/kjfp.2022.29.4.632.

Koley, T. K., Maurya, A., Tripathi, A., Singh, B. K., Singh, M., Bhutia, T. L., Tripathi, P. C., & Singh, B. (2019). Antioxidant potential of commonly consumed underutilized leguminous vegetables. *International Journal of Vegetable Science*, *25*(4), 362–372. doi:10.1080/19315260.2018.1519866.

Kose, B. E., Tacer-Caba, Z., & Nilufer-Erdil, D. (2022). Simulated Digestion of the Pigmented Legumes' (Black Chickpea (*Cicer arietinum* L.) and Brown Lentil (*Lens culinaris* Medikus) Phenolics to Estimate Their Bioavailability. *Plant Foods for Human Nutrition*, *77*(4), 584–590. doi:10.1007/s11130-022-01006-w.

Kumar, A., Singh, N., Kaur, A., & Joshi, R. (2023). Sneak-peek into the chlorophyll content, antioxidant activity, targeted and non-targeted UHPLC-QTOF LC/MS metabolomic fingerprints of pulse microgreens grown under different photoperiod regimes. *Food Bioscience*, *52*, 102506. doi:10.1016/j.fbio.2023.102506.

Lafarga, T., Villaró, S., Bobo, G., Simó, J., & Aguiló-Aguayo, I. (2019). Bioaccessibility and antioxidant activity of phenolic compounds in cooked pulses. *International Journal of Food Science & Technology*, *54*(5), 1816–1823. doi:10.1111/ijfs.14082.

León-López, L., Escobar-Zúñiga, Y., Salazar-Salas, N. Y., Mora Rochín, S., Cuevas-Rodríguez, E. O., Reyes-Moreno, C., & Milán-Carrillo, J. (2020). Improving Polyphenolic Compounds: Antioxidant Activity in Chickpea Sprouts through Elicitation with Hydrogen Peroxide. *Foods*, *9*(12), 1791. doi:10.3390/foods9121791.

López-Fernández, O., Domínguez, R., Pateiro, M., Munekata, P. E. S., Rocchetti, G., & Lorenzo, J. M. (2020). Determination of Polyphenols Using Liquid Chromatography–Tandem Mass Spectrometry Technique (LC–MS/MS): A Review. In *Antioxidants*. 9 (6), 479. doi:10.3390/antiox9060479.

Magalhães, S. C. Q., Taveira, M., Cabrita, A. R. J., Fonseca, A. J. M., Valentão, P., & Andrade, P. B. (2017). European marketable grain legume seeds: Further insight into phenolic compounds profiles. *Food Chemistry*, *215*, 177–184. doi:10.1016/j.foodchem.2016.07.152.

Mahbub, R., Francis, N., Blanchard, ChristopherL., & Santhakumar, AbishekB. (2021). The anti-inflammatory and antioxidant properties of chickpea hull phenolic extracts. *Food Bioscience*, *40*, 100850. doi:10.1016/j.fbio.2020.100850.

Marchiosi, R., dos Santos, W. D., Constantin, R. P., de Lima, R. B., Soares, A. R., Finger-Teixeira, A., Mota, T. R., de Oliveira, D. M., Foletto-Felipe, M. de P., Abrahão, J., & Ferrarese-Filho, O. (2020). Biosynthesis and metabolic actions of simple phenolic acids in plants. *Phytochemistry Reviews*, *19*(4), 865–906. doi:10.1007/s11101-020-09689-2.

Merga, B., & Haji, J. (2019). Economic importance of chickpea: Production, value, and world trade. *Cogent Food & Agriculture*, *5*(1), 1615718. doi:10.1080/23311932.2019.1615718.

Mesfin, N., Belay, A., & Amare, E. (2021). Effect of germination, roasting, and variety on physicochemical, techno-functional, and antioxidant properties of chickpea (*Cicer arietinum* L.) protein isolate powder. *Heliyon*, *7*(9), e08081. doi:10.1016/j.heliyon.2021.e08081.

Milán-Noris, A. K., Gutierrez-Uribe, J. A., & Serna-Saldivar, S. O. (2023). Influence of soaking and boiling on flavonoids and saponins of nine desi chickpea cultivars with potential antiproliferative effects. *Journal of Food Measurement and Characterization*. 17(S1),1-9. doi:10.1007/s11694-023-01861-4.

Munteanu, I. G., & Apetrei, C. (2021). Analytical methods used in determining antioxidant activity: A Review. In *International Journal of Molecular Sciences*, 22 (7), 3380 doi:10.3390/ijms22073380.

Myint, H., Kishi, H., Koike, S., & Kobayashi, Y. (2017). Effect of chickpea husk dietary supplementation on blood and cecal parameters in rats. *Animal Science Journal*, *88*(2), 372–378. doi:10.1111/asj.12651.

Niño-Medina, G., Muy-Rangel, D., de la Garza, A. L., Rubio-Carrasco, W., Pérez-Meza, B., Araujo-Chapa, A. P., Gutiérrez-Álvarez, K. A., & Urías-Orona, V. (2019). Dietary fiber from chickpea (*Cicer arietinum*) and soybean (Glycine max) husk byproducts as baking additives: Functional and Nutritional Properties. *Molecules*, *24*(5). doi:10.3390/molecules24050991.

Osorio-Tobón, J. F. (2020). Recent advances and comparisons of conventional and alternative extraction techniques of phenolic compounds. *Journal of Food Science and Technology*, *57*(12), 4299–4315. doi:10.1007/s13197-020-04433-2.

Perez-Perez, L. M., Huerta-Ocampo, J. Á., Ruiz-Cruz, S., Cinco-Moroyoqui, F. J., Wong-Corral, F. J., Rascón-Valenzuela, L. A., Robles-García, M. A., González-Vega, R. I., Rosas-Burgos, E. C., Corella-Madueño, M. A., & Del-Toro-Sánchez, C. L. (2021). Evaluation of quality, antioxidant capacity, and digestibility of chickpea (*Cicer arietinum* L. cv Blanoro) stored under $N_2$ and $CO_2$ atmospheres. *Molecules*, 26(9), 2773. doi:10.3390/molecules26092773.

Rachwa-Rosiak, D., Nebesny, E., & Budryn, G. (2015). Chickpeas—composition, nutritional value, health benefits, application to bread and snacks: A Review. *Critical Reviews in Food Science and Nutrition*, *55*(8), 1137–1145. doi:10.1080/10408398.2012.687418.

Roorkiwal, M., Bharadwaj, C., Barmukh, R., Dixit, G. P., Thudi, M., Gaur, P. M., Chaturvedi, S. K., Fikre, A., Hamwieh, A., Kumar, S., Sachdeva, S., Ojiewo, C. O., Tar'an, B., Wordofa, N. G., Singh, N. P., Siddique, K. H. M., & Varshney, R. K. (2020). Integrating genomics for chickpea improvement: achievements and opportunities. *Theoretical and Applied Genetics*, *133*(5), 1703–1720. doi:10.1007/s00122-020-03584-2.

Saleh, H. M., Hassan, A. A., Mansour, E. H., Fahmy, H. A., & El-Bedawey, A. E.-F. A. (2019). Melatonin, phenolics content and antioxidant activity of germinated selected legumes and their fractions. *Journal of the Saudi Society of Agricultural Sciences*, *18*(3), 294–301. doi:10.1016/j.jssas.2017.09.001.

Sánchez-Chino, X. M., Jiménez-Martínez, C., Vásquez-Garzón, V. R., Álvarez-González, I., Villa-Treviño, S., Madrigal-Bujaidar, E., Dávila-Ortiz, G., & Baltiérrez-Hoyos, R. (2017). Cooked chickpea consumption inhibits colon carcinogenesis in mice induced with azoxymethane and dextran sulfate sodium. *Journal of the American College of Nutrition*, *36*(5), 391–398. doi:10.1080/07315724.2017.1297744.

Sandeep, D., Kalyan Reddy, C. Y., Reddy, P. Y., & Shafeen, S. (2012). Anti-inflammatory activity of *Cicer arietinum* seed extracts.*Asian Journal of Pharmaceutical and Clinical Research*. https://www.researchgate.net/publication/287535164.

Segev, A., Badani, H., Kapulnik, Y., Shomer, I., Oren-Shamir, M., & Galili, S. (2010). Determination of polyphenols, flavonoids, and antioxidant capacity in colored chickpea (*Cicer arietinum* L.). *Journal of Food Science*, *75*(2), S115–S119. doi:10.1111/j.1750-3841.2009.01477.x.

Serventi, L., & Dsouza, L. V. (2020). Bioactives in Legumes. In L. Serventi (Ed.), *Upcycling Legume Water: from wastewater to food ingredients* (pp. 139–153). Springer International Publishing. doi:10.1007/978-3-030-42468-8_10.

Shang, X., Dou, Y., Zhang, Y., Tan, J.-N., Liu, X., & Zhang, Z. (2019). Tailor-made natural deep eutectic solvents for green extraction of isoflavones from chickpea (*Cicer arietinum* L.) sprouts. *Industrial Crops and Products*, *140*, 111724. doi:10.1016/j.indcrop.2019.111724.

Sharma, A. (2021). A review on traditional technology and safety challenges with regard to antinutrients in legume foods. *Journal of Food Science and Technology*, *58*(8), 2863–2883. doi:10.1007/s13197-020-04883-8.

Siddiqui, N., Ali, T. M., Butt, N. A., Abbas, T., & Hasnain, A. (2022). Effects of onion peel extract on bioactive components, textural and sensory characteristics of chickpea-based fried snack. *Legume Science*, *n/a*(n/a), e163. doi:10.1002/leg3.163.

Singh, B., Singh, J. P., Kaur, A., & Singh, N. (2017). Phenolic composition and antioxidant potential of grain legume seeds: A review. *Food Research International*, *101*, 1–16. doi:10.1016/j.foodres.2017.09.026.

Singh Sibian, M., & Singh Riar, C. (2021). Optimization and evaluation of composite flour cookies prepared from germinated triticale, kidney bean, and chickpea. *Journal of Food Processing and Preservation*, *45*(1), e14996. doi:10.1111/jfpp.14996.

Sofi, S. A., Singh, J., Muzaffar, K., Mir, S. A., & Dar, B. N. (2020). Effect of germination time on physico-chemical, functional, pasting, rheology and electrophoretic characteristics of chickpea flour. *Journal of Food Measurement and Characterization*, *14*(5), 2380–2392. doi:10.1007/s11694-020-00485-2.

Summo, C., De Angelis, D., Ricciardi, L., Caponio, F., Lotti, C., Pavan, S., & Pasqualone, A. (2019). Nutritional, physico-chemical and functional characterization of a global chickpea collection. *Journal of Food Composition and Analysis*, *84*, 103306. doi:10.1016/j.jfca.2019.103306.

Torra, M., Belorio, M., Ayuso, M., Carocho, M., Ferreira, I. C. F. R., Barros, L., & Gómez, M. (2021). Chickpea and chestnut flours as non-gluten alternatives in cookies. *Foods*, *10*(5). doi:10.3390/foods10050911.

Vo, Q. V, Nam, P. C., Thong, N. M., Trung, N. T., Phan, C.-T. D., & Mechler, A. (2019). Antioxidant motifs in flavonoids: O–H versus C–H bond dissociation. *ACS Omega*, *4*(5), 8935–8942. doi:10.1021/acsomega.9b00677.

Wang, J., Li, Y., Li, A., Liu, R. H., Gao, X., Li, D., Kou, X., & Xue, Z. (2021). Nutritional constituent and health benefits of chickpea (*Cicer arietinum* L.): A review. *Food Research International*, *150*, 110790. doi:10.1016/j.foodres.2021.110790.

Wojtunik-Kulesza, K., Oniszczuk, A., Oniszczuk, T., Combrzyński, M., Nowakowska, D., & Matwijczuk, A. (2020). Influence of in vitro digestion on composition, bioaccessibility and antioxidant activity of food polyphenols—a non-systematic review. *Nutrients*, *12*(5), 1401. doi:10.3390/nu12051401.

Xiao, S., Li, Z., Zhou, K., & Fu, Y. (2023). Chemical composition of kabuli and desi chickpea (*Cicer arietinum* L.) cultivars grown in Xinjiang, China. *Food Science & Nutrition*, *11*(1), 236–248. doi:10.1002/fsn3.3056.

Xu, B. J., & Chang, S. K. C. (2007). A comparative study on phenolic profiles and antioxidant activities of legumes as affected by extraction solvents. *Journal of Food Science*, *72*(2), S159–S166. doi:10.1111/j.1750-3841.2006.00260.x.

Xu, B. J., Yuan, S. H., & Chang, S. K. C. (2007). Comparative analyses of phenolic composition, antioxidant capacity, and color of cool season legumes and other selected food legumes. *Journal of Food Science*, *72*(2), S167–S177. doi:10.1111/j.1750-3841.2006.00261.x.

Yegrem, L. (2021). Nutritional composition, antinutritional factors, and utilization trends of Ethiopian chickpea (*Cicer arietinum* L.). *International Journal of Food Science*, *2021*, 1–10. doi:10.1155/2021/5570753.

## Chapter 10

# Composition and Characteristics of Non-Conventional Legumes

**S. González-Cuna[1]**
**M. S. Cid-Gallegos[2]**
**M. F. Juárez-Chairez[3]**
**A. E. Cedillo-Olivos[1]**
**A. Linares-Castañeda[1]**
**and C. Jiménez-Martínez[1,*]**
[1]National School of Biological Sciences, National Polytechnic Institute, Mexico City, Mexico
[2]The Interdisciplinary Professional Unit of Biotechnology, National Polytechnic Institute, Mexico City, Mexico
[3]Laboratory of Immunology and Molecular Biology, Academic Unit of Biological Sciences (UACB), The Autonomous University of Zacatecas, Zacatecas, Mexico

## Abstract

Legumes have recently increased in popularity due to their nutritional quality and eco-friendly production, emitting fewer polluting gases than other protein sources. In 2019, the UN General Assembly proclaimed February 10th as World Legume Day, raising public awareness about legumes' nutritional and environmental benefits. Legumes such as common beans, chickpeas, lentils, and peas are grown and consumed worldwide while others are grown and consumed locally. Hence, this chapter aims to identify lesser-known legume species, specifically *Vigna*

* Corresponding Author's Email: crisjm_99@yahoo.com

In: Legumes: Beyond Their Nutritional Value
Editors: C. Jiménez-Martínez, M. F. Juárez-Chairez and M. S. Cid-Gallegos
ISBN: 979-8-89530-440-2

*unguiculata*, *Pithecellobium*, *Leucaena*, *Hymenaea*, *Canavalia ensiformis*, and evaluate their chemical compositions and potential uses.

**Keywords:** non-conventional legumes, chemical composition, consumption

## 1. Introduction

Rural communities in Mexico and other countries commonly gather one of their main sources of food—comestible plants—in surrounding primary and secondary forests. Legumes are among these comestible plants, and they are an important source of genetic resources. With the current changes in climate and increase in food scarcity, protection of these legumes is of the utmost importance (Biswas & & Mohammed Rahmatullah, 2011). Although these legumes possess high nutritional value, they are only popular locally; they require greater promotion to attract the interest of agriculturists. In other words, they are less well-known compared to common beans, chickpeas, and lentils, even though these lesser-studied and -consumed legumes offer valuable compounds that can meet a population's nutritional needs (Aguilera et al., 2013).

Moreover, these lesser-known legumes, which have limited scientific documentation, have been largely forgotten in the traditional knowledge of many communities due to factors such as the erosion of local culture and ecosystem transformation (Shahrajabian et al., 2021). Therefore, it is crucial for local communities to collaborate with scientific, health, and ecological experts by sharing their ancestral knowledge about the use of non-conventional legumes in their surroundings. Such cooperation can help revive interest in these legumes and promote their wider dissemination (Sutjaritjai, 2022). Thus, this chapter will describe each chemical component found in non-conventional legumes such as *Vigna unguiculata*, *Pithecellobium*, *Leucaena*, *Hymenaea*, and *Canavalia ensiformis*.

## 2. *Vigna Unguiculata*

### 2.1. Generalities

*Vigna unguiculata* belongs to the Fabaceae family, *Vigna* genus, and is commonly known as cowpea. This legume is one of the most ancient crops known to man (Kalloo & Bergh, 2012). It is a versatile, vascular, annual, and warm-season legume. *Vigna* grows in plain foliage, with 25 to 35 °C temperatures and annual rainfall from 750 to 1100 mm (Drahansky et al., 2016). Cowpeas originated in Africa, where, based on archeological evidence in Ghana and domesticated cowpeas, Eastern and Western Africa have been proposed as the center of domestication (Figure 1). Then, it was introduced in Europe around 300 B.C., Asia in 200 B.C., and South America in the 17th century (Herniter et al., 2020). Today, cowpeas' primary production and consumption is in sub-Saharan Africa (SSA), specifically in West and Central Africa. Nigeria is the primary producer of *Vigna* seeds annually, with approximately 2.14 million metric tons, consuming approximately 3 million metric tons (Boukar et al., 2019).

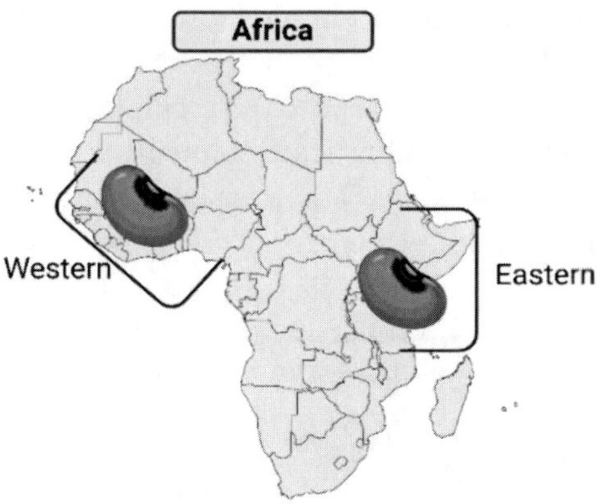


**Figure 1.** Center of domestication of cowpea (Herniter et al., 2020). Created in Biorender.

Cowpeas have cultural importance, as seen in some cultures' traditional food used to celebrate the New Year. For instance, South America has a dish called "Hoppin John," which consists of cowpeas, rice, and bacon. The origin of this tradition could be from Sephardic Jews, who have a similar tradition of preparing the dish to celebrate the Jewish New Year (Herniter et al., 2020).

## 2.2. Chemical Composition

Cowpea's nutritional and nutraceutical properties position it as a vital pulse crop for food security in developing countries, where it serves as one of the primary sources of calories. Further, its high nutrient content makes it a valuable resource for addressing dietary needs and supporting the health of populations in these regions. *Vigna* is essential for its high protein (23-32%), carbohydrate (50-60%), low fat (1%), phytochemical, mineral, and vitamin contents (Table 1), so consuming this seed has a high chance of benefitting the consumer's health (Osipitan et al., 2021).

Carbohydrates are the most significant component in the seed, and are mostly found in starch, a complex carbohydrate. The main sugars in cowpeas are stachyose (3.43%), sucrose (2.97%), and raffinose (1.24%) (Ashogbon & Akintayo, 2013). Moreover, cowpeas contain essential amounts of resistant starch and a higher ratio of slow-digestible to rapid-digestible starch. Resistant starch reduces glycemic response, benefiting people with insulin resistance and diabetes (Ratnaningsih et al., 2016). Finally, α-galactosides present in cowpeas are recognized for their prebiotic function (Thirunathan & Manickavasagan, 2019).

Cowpea's protein content depends on climate factors, though it is approximately 23-32%. Its significant protein fractions are globulins, albumins, glutelins, and prolamins in smaller amounts. Additionally, its amino acid profile is high in lysine, leucine, and arginine (Abebe & Alemayehu, 2022).

The lipid content in cowpea is between 0.5-3.9%, with a distribution of triglycerides (41.2%), phospholipids (25.1%), monoglycerides (10.6%), free fatty acids (7.9%), diglycerides (7.8%), sterols (5.5%), hydrocarbons (2.6%), and sterol esters (5.5%) (Abebe & Alemayehu, 2022). Micronutrients, such as vitamins A and C, are significantly present in cowpeas; thiamin, riboflavin, niacin, vitamin B6, and pantothenic acid are in appreciable quantities. In addition, cowpea is rich in minerals such as potassium, calcium, magnesium, and phosphorus and in small amounts of iron, sodium, zinc, copper, manganese, and selenium (Asare et al., 2013).

## 3. *Pithecellobium*

### 3.1. Generalities

*Pithecellobium* is a legume of a flowering tree that belongs to the Fabaceae family of American origin (Bagchi & Jayaram Kumar, 2016; Murugesan et al., 2019). Usually, it is planted as a living fence or used as food, forage, or firewood. Its pods contain a sweet and acidic pulp that is commonly white in color yet sometimes red. In countries where *Pithecellobium* is harvested, like Mexico, Cuba, and Thailand, pods are sold at roadside stands to be eaten raw or used in beverages similar to lemonade (Bobade, 2019). In countries like India, it is used in curries (Sugumaran et al., 2008). Folk medicine has used *Pithecellobium* for its abortifacient, anodyne, astringent, and larvicidal properties, and to treat convulsions, dysentery, dyspepsia, earache, leprosy, peptic ulcers, sores, toothache, and venereal disease (Bobade, 2019). Likewise, they *have been* used for hundreds of years in Ayurvedic medicine without reported toxicity.

### 3.2. Chemical Composition

*Pitecellobium* is a non-conventional legume distributed in America and Asia. *Pitecellobium's* primary macronutrient is carbohydrates; its secondary macronutrients include proteins, a minimum amount of lipids (see Table 1), and vitamins and minerals (Dhanisha et al., 2022; Ponmozhi et al., 2011). *Pithecellobium* has an essential content of vitamins such as ascorbic acid (133mg), thiamine (0.24 mg), riboflavin (0.1 mg), niacin (0.6 mg), vitamin A (25 UI), and minerals like phosphorous (42 mg), iron (0.5 mg), potassium (20.2 mg), calcium (13 mg), and sodium (19mg) (Ponmozhi et al., 2011).

## 4. *Leucaena*

### 4.1. Generalities

*Leucaena* is a small tree, called "the miracle tree," that grows in tropical and subtropical eco-climatic zones far from frosted zones. Warm temperatures of 25-30°C are needed for optimal tree growth (Alemán-Ramírez et al., 2022). *Leucaena* can be found in Central America, Asia, Africa, and Australia, and there are small distributions in Europe; however, it is absent in Atlantic countries (Nwokocha & Williams, 2012; Ramírez-Bahena et al., 2020). Moreover, in the 80s and 90s, *Leucaena* was introduced to China for reforestation. Some uses reported for *Leucaena* include reforestation, forage for domestic animals, pulp production, woods production, and folk medicine (Ramírez-Bahena et al., 2020).

### 4.2. Chemical Composition

*Leucaena* belongs to the Fabaceae family, which is characterized by rapid growth. *Leucaena* has an essential content of carbohydrates (42.56%), proteins (20-27%), and lipids (6.4-7.5%) (Table 1), as well as vitamins and minerals (Alemán-Ramírez et al., 2022). *Leucaena* has essential amino acids like arginine, alanine, cysteine, glutamic acid, isoleucine, leucine, lysine, and methionine. Further, the majority of *Leucaena*'s lipid content comprises of unsaturated linoleic and oleic fatty acids (72-74%) (Alemán-Ramírez et al., 2022).

**Table 1.** Macronutrients and micronutrients of non-conventional legumes

| Ref.<br><br>Legumes | Abebe & Alemayehu, 2022 | Dhanisha et al., 2022, Sneha et al., 2020 | Alemán-Ramírez et al., 2022 | SiBBr 2018 | Doss et al., 2011 |
|---|---|---|---|---|---|
| Nutrients | *Vigna unguiculata*[1] | *Pithecellobium*[2] | *Leucaena* | *Hymenaea (pulp)*[2] | *Canavalia ensiformis*[2] |
| Carbohydrates | 60.03 | 18.2-19.6 | 42.56% | 33.16 | 50.8 |
| Lipids | 1.26 | 0.4-0.5 | 6.4-7.5% | 2.65 | 4.2 |
| Protein | 23.52 | 2.3-3 | 20–27% | 6.13 | 29.8 |
| Total Dietary Fiber | 10.60 | Nd | Nd | 45.7 | Nd |
| Total Sugar | 6.90 | Nd | Nd | Nd | Nd |
| Ash | 3.24 | 6% | 19.5% | 3.73 | 4.48 |
| Water | 11.95 | 77.8% | Nd | 8.6 | 8.41 |

[1] Raw seed average value percentage of dry matter; [2] g/100g sample,
Nd: Not determined.

## 5. *Hymenaea*

### 5.1. Generalities

Genus *Hymenaea* seeds are known as "jatoba," "jetai," "jatai-uva" and "jetaiba" and have a Neotropical distribution in Central and South America, including one species in East Africa (Oliveira et al., 2012). One of the main characteristics of this species is its content of diterpenoids in bark extracts and resin (Doménech-Carbó et al., 2009). In addition, *Hymenaea* has some biological applications to treat inflammation, bacterial infections, rheumatism, and anemia. Indeed, folk medicine has used *Hymenaea* to treat stomach disorders and bronchitis (Gazzaneo et al., 2005; Agra et al., 2008).

### 5.2. Chemical Composition

*Hymenaea*'s nutritional components have great pharmaceutical potential but are most often used for food. *Hymenaea* mainly comprises of protein (6.13 g), carbohydrates (33.16 g), and fat (2.65 g) (Table 1). Additionally, it contains several minerals, including calcium (74 mg), iron (1.1 mg), magnesium (48 mg) and zinc (1 mg), as well as vitamin A (32 µg) and vitamin C (4.65 mg) (SiBBr, 2018).

## 6. *Canavalia Ensiformis*

### 6.1. Generalities

*Canavalia ensiformis,* generally called jack bean, is the most common tree on the North Coast of Columbia and is a popular ornamental plant in Nigeria (Agbede & Aletor, 2005). It belongs to the *Papilionaceae* family and can effectively control spear grass. Additionally, it serves as a valuable forage resource for livestock (Akpapunam & Sefa-Dedeh, 1997). Although these legumes have several antinutritional components, such as saponins, terpenoids, trypsin inhibitors, and cyanogenic glycosides that limit nutritional availability, thermal treatments—such as boiling, roasting, or steaming— can help to decrease the components, making the nutrients more accessible (Agbede & Aletor, 2005).

### 6.2. Chemical Composition

*Canavalia ensiformis* is an underutilized legume accessible in Latin America, Africa, Asia, the West Indies, Indonesia, and India (Munjal et al., 2024). In addition, Canavalia is leguminous and rich in protein like *Phaseolus vulgaris*. Its primary component is carbohydrates, and its lipid content is low (Abitogun & Olasehinde, 2012). On the other hand, Canavalia has a good concentration of minerals: sodium (2.53 mg/100g); potassium (5.92 mg/100g); calcium (3.21 mg/100g); magnesium (1.95 mg/100g); phosphorus (1.5 mg/100g); and zinc (2.90 mg/100g) (Abitogun & Olasehinde, 2012).

## Conclusion

Non-conventional legumes are a promising biological resource that have been used in folk medicine for many years and have an essential nutritional composition. Their dietary components can be used to prevent or cure some diseases and can also be applied in the food and materials industries, depending on their techno-functional properties. However, further research is needed to better understand and utilize local legume species.

## References

Abebe, B. K., & Alemayehu, M. T. (2022). A review of the nutritional use of cowpea (*Vigna unguiculata* L. Walp) for human and animal diets. *Journal of Agriculture and Food Research, 10,* 100383. https://doi.org/https://doi.org/10.1016/j.jafr.2022.100383.

Abitogun, A., & Olasehinde, E. (2012). Nutritional evaluation of seed and characterization of crude jack bean (*Canavalia ensiformis*) oil. *Journal of Applied Chemistry, 1*(6), 36-40.

Agbede, J. O., & Aletor, V. A. (2005). Studies of the chemical composition and protein quality evaluation of differently processed *Canavalia ensiformis* and *Mucuna pruriens* seed flours. *Journal of Food Composition and Analysis, 18*(1), 89-103. https://doi.org/https://doi.org/10.1016/j.jfca.2003.10.011.

Agra, M. d. F., Silva, K. N., Basílio, I. J. L. D., Freitas, P. F. d., & Barbosa-Filho, J. M. (2008). Survey of medicinal plants used in the region Northeast of Brazil. *Revista brasileira de farmacognosia, 18,* 472-508. https://doi.org/ https://doi.org/10.1590/S0102-695X2007000100021.

Aguilera, Y., Díaz, M. F., Jiménez, T., Benítez, V., Herrera, T., Cuadrado, C., Martín-Pedrosa, M., & Martín-Cabrejas, M. A. (2013). Changes in nonnutritional factors and antioxidant activity during germination of nonconventional legumes. *Journal of agricultural and food chemistry, 61*(34), 8120-8125. https://doi.org/10.1021/jf4022652.

Akpapunam, M. A., & Sefa-Dedeh, S. (1997). Jack bean (*Canavalia ensiformis*): Nutrition related aspects and needed nutrition research. *Plant Foods for Human Nutrition, 50*(2), 93-99. https://doi.org/10.1007/BF02436029.

Alemán-Ramirez, J. L., Okoye, P. U., Torres-Arellano, S., Mejía-Lopez, M., & Sebastian, P. J. (2022). A review on bioenergetic applications of *Leucaena leucocephala. Industrial Crops and Products, 182,* 114847. https://doi.org/https://doi.org/10.1016/j.indcrop.2022.114847.

Asare, A. T., Agbemafle, R., Adukpo, G. E., Diabor, E., & Adamtey, K. A. (2013). Assessment of functional properties and nutritional composition of some cowpea (*Vigna unguiculata* L.) genotypes in Ghana. *Journal of Agricultural and Biological Science, 8*(6), 465-469.

Ashogbon, A., & Akintayo, E. (2013). Isolation and characterization of starches from two cowpea (*Vigna unguiculata*) cultivars. *International Food Research Journal*, *20*(6).

Bagchi, S., & Jayaram Kumar, K. (2016). Studies on water soluble polysaccharides from *Pithecellobium dulce* (Roxb.) Benth. seeds. *Carbohydrate Polymers*, *138*, 215-221. https://doi.org/https://doi.org/10.1016/j.carbpol.2015.11.018.

Biswas, K. R., & Mohammed Rahmatullah, M. R. (2011). A survey of non-conventional plants consumed during times of food scarcity in three adjoining villages of Narail and Jessore Districts, Bangladesh. *American-Eurasian Journal of Sustainable Agriculture*, *5*(1), 1-5.

Bobade, A. F. (2019). GC-MS Analysis of bioactive compound in ethanolic extract of *Pithecellobium dulce* leaves. *Acta Scientific Pharmaceutical Sciences*, *3*(11), 08-13. https://doi.org/10.31080/ASPS.2019.03.0412.

Boukar, O., Belko, N., Chamarthi, S., Togola, A., Batieno, J., Owusu, E., Haruna, M., Diallo, S., Umar, M. L., Olufajo, O., & Fatokun, C. (2019). Cowpea (*Vigna unguiculata*): Genetics, genomics and breeding. *Plant Breeding*, *138*(4), 415-424. https://doi.org/https://doi.org/10.1111/pbr.12589.

Dhanisha, S. S., Drishya, S., & Guruvayoorappan, C. (2022). *Pithecellobium dulce* induces apoptosis and reduce tumor burden in experimental animals via regulating pro-inflammatory cytokines and anti-apoptotic gene expression. *Food and Chemical Toxicology*, *161*, 112816. https://doi.org/https://doi.org/10.1016/j.fct.2022.112816.

Doménech-Carbó, M. T., de la Cruz-Cañizares, J., Osete-Cortina, L., Doménech-Carbó, A., & David, H. (2009). Ageing behaviour and analytical characterization of the Jatobá resin collected from Hymenaea stigonocarpa Mart. *International Journal of Mass Spectrometry*, *284*(1), 81-92. https://doi.org/https://doi.org/10.1016/j.ijms.2008.12.015.

Doss, A., Pugalenthi, M., Vadivel, V., Subhashini, G., & Subash, A. R. (2011). Effects of processing technique on the nutritional composition and antinutrients content of under--utilized food legume *Canavalia ensiformis* L. DC. *International Food Research Journal*, *18*(3).

Drahansky, M., Paridah, M., Moradbak, A., Mohamed, A., Owolabi, F., Asniza, M., & Abdul-Khalid, S. (2016). Cowpea: a strategic legume species for food security and health. In Intechopen (Ed.), *Legume Seed Nutraceutical Research Dietary* (pp. 47-65). Intechopen. https://doi.org/10.5772/intechopen.79006.

Gazzaneo, L. R. S., De Lucena, R. F. P., & de Albuquerque, U. P. (2005). Knowledge and use of medicinal plants by local specialists in an region of Atlantic Forest in the state of Pernambuco (Northeastern Brazil). *Journal of Ethnobiology and Ethnomedicine*, *1*, 1-8.

Herniter, I. A., Muñoz-Amatriaín, M., & Close, T. J. (2020). Genetic, textual, and archeological evidence of the historical global spread of cowpea ([L.] Walp.). *Legume Science*, *2*(4), e57. https://doi.org/https://doi.org/10.1002/leg3.57.

Kalloo, G., & Bergh, B. (2012). *Genetic improvement of vegetable crops*. Newnes.

Munjal, S. D., Dhankhar, J., Sharma, A., & Dhull, S. B. (2024). Physicochemical properties of native Jack bean (*Canavalia ensiformis*) starch: An underutilised legume. *Journal of Applied and Natural Science*, *16*(1), 410-419.

Murugesan, S., Lakshmanan, D. K., Arumugam, V., & Alexander, R. A. (2019). Nutritional and therapeutic benefits of medicinal plant *Pithecellobium dulce* (Fabaceae): A review. *Journal of Applied Pharmaceutical Science*, *9*(7), 130-139.

Nwokocha, L. M., & Williams, P. A. (2012). Rheological characterization of the galactomannan from *Leucaena leucocephala* seed. *Carbohydrate Polymers*, *90*(2), 833-838. https://doi.org/https://doi.org/10.1016/j.carbpol.2012.06.008.

Oliveira, V. B., Yamada, L. T., Fagg, C. W., & Brandão, M. G. L. (2012). Native foods from Brazilian biodiversity as a source of bioactive compounds. *Food Research International*, *48*(1), 170-179. https://doi.org/https://doi.org/10.1016/j.foodres. 2012.03.011.

Osipitan, O. A., Fields, J. S., Lo, S., & Cuvaca, I. (2021). Production Systems and Prospects of Cowpea (*Vigna unguiculata* (L.) Walp.) in the United States. *Agronomy*, *11*(11).

Ponmozhi, P., Geetha, M., Kumar, S., & Devi, S. (2011). Extraction of anthocyanin and analysing its antioxidant properties from *Pithecellobium dulce* fruit pericarp. *Asian journal of pharmaceutical and clinical research*, *4*(1), 41-45.

Ramírez-Bahena, M. H., Flores-Félix, J. D., Velázquez, E., & Peix, Á. (2020). The Mimosoid tree *Leucaena leucocephala* can be nodulated by the symbiovar genistearum of *Bradyrhizobium canariense*. *Systematic Applied Microbiology*, *43*(1), 126041. https://doi.org/10.1016/j.syapm.2019.126041.

Ratnaningsih, N., Harmayani, E., & Marsono, Y. (2016). Composition, microstructure, and physicochemical properties of starches from Indonesian cowpea (*Vigna unguiculata*) varieties. *International Food Research Journal*, *23*(5).

Shahrajabian, M. H., Sun, W., & Cheng, Q. (2021). Bioactive Components and Chemical Constituents of Some Important Legumes in Traditional Medicine. *Journal of Stress Physiology & Biochemistry*, *17*(4), 30-45.

SiBBr, S. d. I. s. a. B. B. (2018). *Hymenaea stigonocarpa*. Biodiversity for food and nutrition. Retrieved May 20 from http://www.b4fn.org/resources/species-database/detail/hymenaea-stigonocarpa/.

Sneha, D., Prashanth, S., Kaveti, V. S., & Boggula, N. (2020). Systematic Review of *Pithecellobium Dulce* (Roxb.) Benth.: A Traditional Medicinal Herb. *Journal for Innovative Development in Pharmaceutical and Technical Science (JIDPTS)*, *3*(5), 1-9.

Sugumaran, M., Vetrichelvan, T., & Quine, S. D. (2008). Free radical scavenging activity of folklore: *Pithecellobium dulce* Benth. leaves. *Ethnobotanical Leaflets*, *2008*(1), 54.

Sutjaritjai, N., Panyadee, P., Phumthum, M., Inta, A., & Balslev, H. (2022). High Diversity of Medicinal Uses of Thai Legumes (Fabaceae) and Their Potential in Public Herbal Medicine. *Diversity*, *14*(8).

Thirunathan, P., & Manickavasagan, A. (2019). Processing methods for reducing alpha-galactosides in pulses. *Critical reviews in food science and nutrition*, *59*(20), 3334-3348. https://doi.org/10.1080/10408398.2018.1490886.

## Chapter 11

# The Role of *Pisum sativum* in Human Health

**V. Pérez-Pérez**[1]
**S. E. García-Solís**[2]
**C. P. Plazola-Jacinto**[2]
**M. Nicolas-García**[3]
**L. Alamilla-Beltrán**[2]
**and L. J. Corzo-Ríos**[1,*]

[1]Departamento de Bioprocesos, Unidad Profesional Interdisciplinaria de Biotecnología, Instituto Politécnico Nacional, México City, México
[2]Departamento de Ingeniería Bioquímica, Escuela Nacional de Ciencias Biológicas, Instituto Politécnico Nacional (IPN), Unidad Profesional Adolfo López Mateos, Zacatenco, Mexico City, Mexico
[3]Tecnológico Nacional de México, Instituto Tecnológico Superior de Teziutlán, División de Ingeniería en Industrias Alimentarias, Teziutlán, Puebla, Mexico

### Abstract

The pea (*Pisum sativum* L.) is a widely consumed legume that has been the subject of recent research due to its health benefits. The dry pea is a rich source of low-digestible carbohydrates, starch, protein, vitamins, minerals, and phytochemicals. Low-digestible carbohydrates help improve gastrointestinal function and reduce starch digestibility. Pea protein is recognized for its high quality, with numerous studies emphasizing its health benefits, including antioxidant and antihypertensive properties, as well as its potential to modulate intestinal microbiota activity. When hydrolyzed, the pea protein may also yield

* Corresponding Author's Email: lcorzo@ipn.mx

In: Legumes: Beyond Their Nutritional Value
Editors: C. Jiménez-Martínez, M. F. Juárez-Chairez and M. S. Cid-Gallegos
ISBN: 979-8-89530-440-2

peptides with bioactivities, including the angiotensin I-converting enzyme inhibitor activity. Peas are also rich in vitamins and minerals, vital in preventing nutritional-deficiency-related diseases, particularly those related to Se or folate deficiencies. The phytochemical composition of pea plants has received considerable attention, and various phenolic principles, such as flavonoids, isoflavonoids, phenolic acids, and other minor phenolics and phytoalexins, have been characterized, which may have antioxidant and anticarcinogenic activity. Saponins, which may exhibit hypocholesterolemic and anticarcinogenic activity, and galactose oligosaccharides, which may exert beneficial prebiotic effects in the large intestine, are other constituents that have been identified. This chapter reviews the current research on the health benefits of bioactive compounds and micro and macronutrients provided by pea consumption.

**Keywords**: pea, bioactive compounds, biological activities, human health

## 1. Introduction

*Pisum sativum* L., commonly known as pea, is a self-pollinated diploid annual crop belonging to the legume family. *Pisum genus* comprises the third largest family of flowering plants with more than 650 genera and 1800 species. This legume is one of the main sources of human food, consumed as a fresh, dehydrated, canned, or frozen vegetable. Peas rank fourth in world crop production behind soybeans, corn, and beans. It is grown on 10.7 million hectares, and approximately 37 million tons are produced, mainly in Canada, Russia, China, India, and the United States of America (Khan et al., 2013; FAOSTAT 2019). An increase of approximately 36% in dry pea production has been seen from 2007 to date, mainly in countries such as Ireland, Belgium, the Netherlands, and Denmark (FAOSTAT 2019).

Peas have been widely cultivated for centuries and are recognized as an inexpensive and accessible source of nutrients. The crop thrives in cold climates with temperatures ranging from 6 to 30°C and prefers light-textured, slightly acidic soils. However, peas are sensitive to high calcium levels, which can lead to chlorosis and grain hardening (Parihar et al., 2020). It is an annual climbing herbaceous plant with hollow stems that grow up to 2 to 3 m tall, with compound pinnate leaves that are 1.5 to 8 cm long, and white to purple flowers consisting of five petals of different sizes. The fruit grows in a pod 2.5 to 10 cm long formed by two sealed valves that have a rough inner membrane in which the seeds are contained, the seeds are round, smooth, and green in color (Rungruangmaitree and Jiraungkoorskul, 2017; Zilani et al., 2017).

The genus *Pisum* L. has mainly three species (*P. sativum* subsp. *sativum*) with its five subspecies (*elatius, sativum, humile, arvense*, and *hortense*), including green peas, garden peas, field peas, whole pod edible peas, and wild peas. In the human diet, peas are consumed as fresh vegetables, in soups, and snacks, and their flour is also often used to manufacture food products. Because it is an excellent source of protein (21.2–32.9%), starch (36.9-49.0%), and fiber (14-26%), it is being widely used as an ingredient in many food industries around the world (Dahl et al., 2013).

## 2. Nutritional Composition

*Pea protein* is considered one of the cheapest sources of easily digestible protein for human consumption. It is more digestible than other vegetable proteins, and its consumption causes fewer allergic reactions. It is mainly composed of globulins (55-65%) and albumins (18-25%), although it also contains low proportions of prolamins (4-5%) and glutelins (3-4%). It has been reported that pea protein, like the proteins of other legumes, contains a good proportion of lysine but tends to be limited in methionine and tryptophan (Dahl et al., 2013; Lam et al., 2016).

Legume globulins are classified into two main classes: legumin and vicilins. Legumin is a hexameric protein constituted by acid and basic subunits linked by a disulfide bond. The molecular mass of legumin is 300-400 kDa. Vicilins are smaller proteins with 150-170 kDa sizes, containing three monomers linked by hydrophobic interactions. Vicilin contains low levels of sulfur amino acids (methionine, cysteine) and tryptophan and higher levels of basic (arginine, lysine) and acidic (aspartic acid, citric acid) amino acids (Mertens et al., 2012; Lam et al., 2018). Peas are 50-60% globulins and 12-15% albumins, and the two major globulins are 7S and 11S named vicilin and convicilin, respectively (Rubio et al., 2014; Djoullah et al. 2018). The ratio of albumins to globulins in pea protein depends on the species and the production methods employed. These differences can affect the functional properties of the pea protein. An increase in the concentration of globulins can improve the protein emulsification capacity. In addition to the functional properties of pea protein, it can produce bioactive peptides that can act as antioxidants and inhibitors of the angiotensin I-converting enzyme (ACE) activity. Therefore, its consumption has positively affected human health (Burger and Zhang, 2019).

Additionally, peas contain up to 65% carbohydrates, mainly starch. Pea starch is composed of amylose (30-60%) and amylopectin (40-70%); the high amylose content of this starch allows it to be used for film formation, representing an advantage over other vegetable starch sources. Another important advantage of this legume is its percentage of resistant starch (2-10%); this fraction is not digested in the intestine. Indigestible carbohydrates, such as resistant starch, cellulose, hemicellulose, and pectin, form an insoluble dietary fiber. Peas contain more of this compound (17-27%) than wheat and other cereals; thus, their consumption can improve gastrointestinal health (Xiaojing et al., 2015; Tulbek et al., 2017). Soluble and insoluble fiber are present in both the seed pericarp and cotyledon of peas. Additionally, the pericarp of this legume contains the highest concentration of micronutrients, including vitamins and minerals (Parihar et al., 2016).

Peas contain higher concentrations of minerals, such as calcium, magnesium, phosphorus, iron, zinc, and copper, than wheat and other cereals. In addition, peas are a rich source of folic acid, niacin, thiamine, riboflavin, pyridoxamine, pyridoxal, and pyridoxine. Therefore, peas may be combined with other cereals to provide a balanced mineral and vitamin concentration that may improve the nutritional health of developing and underdeveloped nations (Tulbek et al., 2017). Non-nutritive compounds that can improve consumers' health are also found in peas. These molecules are bioactive compounds that include phenols, polyphenols, saponins, anthocyanins, flavonoids, and tannins. Antioxidant compounds are molecules that have the function of neutralizing free radicals produced during oxidation processes. Among the phenolic compounds in peas are coumaric, caffeic, and vanillic acids, as well as flavonoids such as kaempferol, daidzein, genistein, asparaginase, apigenin, and lectin (Kumari and Deka, 2021).

Besides peas' nutritional and non-nutritional content, this legume has low concentrations of anti-nutritional compounds compared to other legumes such as soybeans. Anti-nutritional compounds impair the digestibility and functionality of nutrients. Among the main anti-nutritive compounds contained in peas is phytic acid, which binds to metals in the stomach, decreasing the digestibility of proteins, and is found mainly in the peel associated with some proteins. Peas also contain small concentrations of trypsin and chymotrypsin inhibitors, which are small peptides found in the cotyledon that also reduce protein digestibility and bioavailability. Another anti-nutritional compound in peas is a lectin, which agglutinates red blood cells, alters the activity of some enzymes and reduces the viability of epithelial cells. Some strategies to reduce the concentration of anti-nutritional compounds in peas are soaking, fermentation, heat treatment, and germination (Daryanto et al., 2015; De Angelis et al., 2020).

Due to this nutritional composition, incorporating peas into a diet can provide several health benefits; modern pharmacology reveals that peas have a high medicinal value. Regularly eating peas in the daily diet can effectively reduce the occurrence of type II diabetes, colon cancer, and heart disease. The main effects of pea consumption on human health are described below.

## 3. Health Benefits

### 3.1. Antioxidant Activity

Some parts of pea plants have shown remarkable activity against a wide range of free radicals and important metal-chelating activities. Among these, the seeds and pods have been the most extensively studied, primarily for their phenolic content and antiradical activity. These studies have employed various in vitro assays operating through different mechanisms, including DPPH, ABTS, FRAP, TEAC, TAA, beta-carotene linoleate, xanthine-xanthine oxidase, and ORAC models. The literature consistently indicates that, due to a higher concentration of bioactive compounds in the seed coat compared to the cotyledons, the antioxidant scavenging capacity of pea seeds is predominantly attributed to the seed coat (Martens et al., 2017; Fahim et al., 2019).

Peas' antioxidant activity is largely due to its contents of phenolic compounds, carotenoids, anthocyanins, and flavonoids. These compounds have antioxidant, anti-inflammatory, and antihypertensive properties. The bioactive compounds are secondary metabolites the plant produces that help defend against pathogenic microorganisms, parasites, and predators. They include many heterogeneous substances with at least one aromatic ring and a hydroxyl group with a functional side chain. Some of the bioactive compounds reported for peas are shown in Table 1 (Borges-Martínez et al., 2020; Costantini et al., 2021). The hull and cotyledons contain most of their bioactive compounds. Proanthocyanidins, flavonoids, and flavonols have been reported to be the main antioxidants in the hull. The antioxidant properties of the cotyledons have been attributed mainly to the presence of catechins.

However, the variety and environmental conditions under which peas are grown determine their level of bioactive compounds (Dueñas et al., 2006; Fahim et al., 2019).

In addition to the bioactive compounds listed in Table 1, recent studies have shown that some of the peptides obtained from the hydrolysis of pea protein also have antioxidant activity. These peptides can be obtained by different techniques, which include enzymatic hydrolysis, germination, fermentation, physical (high hydrostatic pressure, ultrasounds, microwave-assisted extractions, ohmic heating, pulsed electric fields), and chemical (use of acids or alkalis to hydrolysate protein) methods. The techno-functional properties of the protein to be used in the food industry are improved by these methods (Girgih et al., 2015; Wang et al., 2017; Zha et al., 2019; Li et al., 2020). Hydrolysis of the pea protein produces peptides with a size of less than 3 kDa with a higher concentration of hydrophobic amino acids (alanine, valine, isoleucine, leucine, tyrosine, phenylalanine, tryptophan, proline, methionine, and cysteine) and aromatic amino acids (phenylalanine, tryptophan, and tyrosine). The fractions with high hydrophobic and aromatic amino acid content exhibited the pea protein's strongest radical scavenging and metal-chelating activities (Pownall et al., 2010; Pownall et al., 2011; Ge et al., 2020).

Compounds with antioxidant activity can inhibit or reduce the damage caused by the action of free radicals or reactive non-radical species, preventing oxidative damage. Free radicals are molecules that have unpaired electrons in their outer orbitals and are, therefore, charged and unstable species that can take an electron from another atom through oxidation or give an electron through reduction. These species are produced during respiration, specifically in the electron transport chain. Although most of them reach the pump of the electron transport system, between 1% and 3% react with oxygen to form a superoxide radical. Free radicals help neutrophils and macrophages to destroy bacteria. Still, they can also damage surrounding tissue cells, causing oxidative stress (OS) and inflammation, promoting cell proliferation, and causing alterations in proteins and nucleic acid (Fox, 2017; Adwas et al., 2019). Many diseases, including diabetes, heart disease, atherosclerosis, liver disease, and cancer, develop and progress as a result of the formation of these free radicals (Halliwell, 2007; Valko et al., 2007).

**Table 1**. Bioactive compounds from peas with antioxidant activity

| Bioactive compounds | | | | | |
|---|---|---|---|---|---|
| Phenolic acids and derivatives | Carotenoids | Anthocyanins | Flavonoids and derivatives | Phytosterols | References |
| Gallic acid, syringic acid, ferulic acid. | NR | NR | Quercetin, catechin | NR | Borges-Martínez et al., 2020. |
| Protocatechuic acid hexoside, p-hydroxybenzoic acid, protocatechuic acid, caffeic acid. | NR | NR | (epi)catechin conjugate, (epi)gallocatechin dimer, (epi)gallocatechin trimer, (epi)catechin conjugate, (epi)gallocatechin, quercetin derivative, catechin, quercetin 3-sophorotriose, quercetin dihexoside, kaempferol trihexoside, quercetin 3-rutinoside, quercetin 3-glucoside, quercetin 3-hexoside, quercetin deoxyhexoside, melitidin, quercetin, theaflavin derivative, kaempferol, apigenin, isorhamnetin glycoside, kaempferol hexoside, naringenin. | NR | Guo et al., 2019. |
| Ferulic acid. | β-carotene | Cyanidin 3-O-glucoside | NR | NR | Constantini et al., 2021 |
| ρ-Coumaric acid, ellagic acid. | NR | NR | Kaempferol-3-neohesperidoside, kaempferol-3- glucoside, naringenin. | β-sitosterol | Hossain et al., 2017 |

**Table 1**. (Continued)

| Bioactive compounds | | | | | |
|---|---|---|---|---|---|
| Phenolic acids and derivatives | Carotenoids | Anthocyanins | Flavonoids and derivatives | Phytosterols | References |
| Caffeic acid, chlorogenic acid, cinnamic acid, o-coumaric acid, 2,3-dihydroxybenzoic acid, ferulic acid, gallic acid, salicylic acid, sinapic acid. | NR | Delphinidin 3-O-(2-O-β-D-xylopyranosyl-β-D-galactopyranoside)-5-Oβ-D-glucopyranoside, delphinidin 3-O-(2-O-β-D-xylopyranosyl-β-D-galactopyranoside)-5-O(6-O-acetyl)-β-D-glucopyranoside. | Kaempferol, quercetin. | NR | Fahim et al., 2019 |
| Caffeic acid, coumaric acid, gentisic acid, ferulic acid, protocatechuic acid, vanillic acid, pisatin. | NR | Proanthocyanidin | Apigenin, daidzein, genistein, kaempferol, catechin | NR | Rungruangmaitree and Jiraungkoorskul, 2017 |
| Coumaric acids, caffeic acids, vanillic acids, ferulic, pisatin, | NR | Proanthocyanidin | Daidzein, genistein, asparaginase, apigenin, lectin, kaempferol, catechin | NR | Kumari and Deka, 2021 |

NR = Not reported.

Although the body has a built-in defense mechanism to eliminate free radicals, additional dietary antioxidants are needed to eliminate these compounds and prevent them from causing health damage (Ge et al., 2020; Jideani et al., 2021). Dietary antioxidant compounds help neutralize reactive oxygen species (ROS) produced during OS, restoring balance. The mechanisms by which these bioactive compounds neutralize free radicals depend on the hydroxylation status of their aromatic rings, conjugation with other phenols, degree of glycosylation/acylation, size, and solubility (Ramos, 2008; Ozcan et al., 2014). These mechanisms could include scavenging free radicals, chelation and stabilization of divalent cations, and modulation of endogenous antioxidant enzymes. In addition, bioactive compounds can also act as substrates for radicals such as superoxides and hydroxyl radicals (Adwas et al., 2019).

Phenolic acids and flavonoids protect deoxyribonucleic acid (DNA) from oxidation by inhibiting free radicals and enzymes activating pro-carcinogens. They also activate xenobiotic detoxification enzymes. For example, a synergistic protective effect against DNA oxidation has been reported for L-ascorbic acid and flavonoids. Furthermore, flavonoids and their derivatives, especially catechins, protect the brain against aging and prevent the development of progressive neurodegenerative disorders like Parkinson's and Alzheimer's (Weinreb et al., 2004; Ozcan et al., 2014).

Guo et al. (2019) reported that the administration of a pea phenolic extract to rats reduced the malondialdehyde (MDA) plasma concentration and increased the amount of the glutathione peroxidase (GSH) enzyme. MDA is a product of the polyunsaturated fatty acids peroxidation process. This process depends on the free radicals' concentration, so the reduced levels of oxidation products, such as MDA, may be due to the inactivation of the phenolic compounds in peas. GSH is one of the enzymes involved in ROS conversion and catalyzes the reduction of peroxide or lipoperoxide, using reduced glutathione as the reducing agent. Lipoperoxides generate ROS and are toxic to animal tissues.

## 3.2. Reduction of Type II Diabetes Occurrence

Type II diabetes mellitus is a metabolic disease resulting from a deficiency in the production and effect of insulin, characterized by the accumulation of high glucose concentrations in blood, known as hyperglycemia. Hyperglycaemia is associated with the massive production of free radicals since elevated blood glucose levels are susceptible to oxidation, leading to excessive ROS production, and there is a link between diabetes and OS (Holman et al., 2008; Cade et al., 2008). Some studies indicate that patients with diabetes have high levels of OS biomarkers, like 8-isoPGF2α, which is the major isoprostane generated during the enzymatic peroxidation of arachidonic acid and is a reliable indicator of OS Another indicator used to link OS to diabetes mellitus is glycaemic variability, because high glucose excretion levels during postprandial periods lead to the generation of excess ROS, exceeding the antioxidant capacity (Sampson et al., 2002; Ito et al., 2019).

Reduction in diabetes development is related to consuming foods that lower blood sugar levels (Martens et al., 2017). The consumption of peas can decrease the changes caused by this disease. Mejri et al. (2019) studied the effect of pea methanol extract consumption on the OS markers in diabetic mice. The results of this study suggest that the extract may reduce OS through the activation of antioxidant enzymes and reduction of lipid peroxidation, which may amend diabetes-related health complications. Other mechanisms produced by pea extract administration were increased insulin release through pancreatic β-cell regeneration and activation of insulin-induced lipoprotein hydrolysis.

The effect of pea ethanolic extract administration in mice revealed that treatment with different doses (100 and 200 mg/kg) significantly reduced blood glucose levels by 18.15% and 30.24% 30 minutes after administration. The antioxidant compounds present in peas contribute to reducing OS, and improving glucose tolerance. In the ethanol extract, different compounds were identified, including β-sitosterol. This compound stimulates basal glucose uptake by activating AMP-activated protein kinase (AMPK) mediated by LKB1 (liver kinase B1). Ellagic acid potentiates pancreatic insulin secretion by increasing glucose uptake, decreasing glucose intolerance, and reducing the absorption of this molecule by inhibiting the enzyme glucosidase. The flavonoid kaempferol enhances insulin-stimulated glucose uptake in mature adipocytes, and naringenin suppresses carbohydrate absorption in the intestine and exerts an extra-pancreatic action (Hossain et al., 2017).

Postprandial glucose levels are associated with more long-term negative health effects than fasting glucose levels. Since blood glucose levels are affected by the type and amount of carbohydrates consumed, foods with high concentrations of dietary fiber, which effectively regulate the postprandial glycaemic response by slowing the glucose rate digestion and absorption, are suggested (Brennan, 2005). Lambert et al. (2014) studied the effect of pea husk fiber intake on postprandial blood glucose levels, observing a significant reduction in postprandial blood glucose levels with it (15 g). Additionally, a significant reduction in the level of insulin was observed 30 minutes after the dietary fiber intake. Lunde et al. (2011) reported that consuming pea fibre-enriched bread reduced the postprandial concentration of glucose and prolonged the feeling of satiety.

In another study, foods made from peas and potatoes were tested for their glycaemic response and insulin production. The glycemia observed after consuming 36 g of carbohydrates from dried peas was only about one-third of that observed after consuming 36 g of carbohydrates from potatoes. These results suggest that pea fiber content can reduce the glycaemic index. Therefore, pea use as a dietary supplement for patients with diabetes is recommended (Schäfer et al., 2003).

In addition to consuming foods with antioxidant compounds and fiber, it is recommended to consume protein. According to the American Diabetes Association, protein consumption appears to reduce postprandial glycemia because it increases the uptake of glucose by tissues through the incretin hormones, such as the glucose-dependent insulinotropic peptide (GIP) and the glucagon-like peptide-1 (GLP-1) (American Diabetes Association, 2017). The incretin hormones are gluco-regulatory hormones responsible for more than 70% of postprandial insulin secretion, particularly GLP-1. Furthermore, people with diabetes may have reduced incretin levels due to the destruction of beta cells in the pancreas (Tasyurek et al., 2014). Tan et al. (2017) assessed the effects of adding plant proteins, such as pea protein, to a sugar-sweetened beverage, which could alter postprandial responses of glucose, insulin, triglycerides, GIP, GLP-1, and appetite. The results showed that adding pea protein reduced the glycaemic response and significantly stimulated insulin release.

Moreover, some oligopeptides present in peas also have a hypoglycaemic effect. These oligopeptides are produced during the enzymatic hydrolysis of pea protein. In the study by Wei et al. (2019), the oligopeptide obtained by hydrolysis had a high content of protein (88.3%) and peptides (81.27%). Amino acid composition of the oligopeptide revealed that it contained considerable amounts of glutamic acid (16.19%), aspartic acid (9.14%), leucine (5.44%), lysine (5.26%), arginine (4.18%), and serine (3.94%). Four peptides, namely Ala-Leu-Pro, Leu-Leu-Pro, Val-Leu-Pro, and Ser-Pro, were present in this study, Ser-Pro exhibited dipeptidase-IV (DPP-IV) inhibitory effects. The results of this study reveal that pea oligopeptides can reduce blood glucose levels and improve glucose tolerance, as well as promote insulin release and glycogen synthase, enhancing the effect of metformin. Most of the functions of the oligopeptides are due to their ability to inhibit the activity of the DPP-IV enzyme.

Some studies indicate that peptides with a sequence ranging from 2 to 8 amino acids containing hydrophobic residues such as aspartic acid and glutamic acid can inhibit the activity of DPP-IV. This enzyme breaks down GLP-1, stimulates insulin secretion and biosynthesis, inhibits glucagon secretion, and regulates gastric emptying. However, only 10% of GLP-1 can reach the pancreas as DPP-IV immediately metabolizes it; therefore, through inhibition of the DDP-IV enzyme, the physiological functions of GLP-1 are maintained for a longer time (Sato et al., 2018).

Quin et al. (2019) analyzed three oligopeptides obtained and characterized from peas and evaluated for their hypoglycaemic activity, obtaining and characterizing three fractions called PGP1, PGP2, and PGP3. PGP2 was the oligopeptide with the highest protein content (85.93%), and its molecular structure contains mannose, ribose, galacturonic acid, glucose, galactose, and arabinose, as well as aspartic acid, glutamic acid, cysteine, phenylalanine, and lysine. Pea oligopeptides reduce the activity of the enzyme's alpha-amylase and alpha-glucosidase by reversible, non-competitive inhibition, which is responsible for hydrolyzing sucrose and/or maltose, releasing glucose into the bloodstream and causing postprandial hyperglycemia. PGP2 and PGP3 had a greater effect on enzyme inhibition. However, PGP2 had the best inhibitory effect on glucose uptake, even after the *in vitro* digestion process.

Subsequently, Li et al. (2021) evaluated the hypoglycaemic activity of PGP2 *in vivo.* The results showed that ingestion of PGP2 in mice was able to alleviate weight loss symptoms caused by diabetes, reduce blood sugar levels, improve oral glucose tolerance, promote secretion, improve resistance, and increase insulin sensitivity. In addition to the above, PGP2 showed a protective effect on the organs of diabetic mice. These two studies show that PGP2 may contribute to diabetes prevention and control.

### 3.3. Preventing Cardiovascular Diseases

Cardiovascular diseases currently represent a significant public health concern in many industrialized countries. Their development is closely associated with risk factors such as hypercholesterolemia, hypertension, type II diabetes, sedentary lifestyles, high-fat diets, and stress. Research shows that OS-induced free radicals play a key role in cellular processes involved in atherosclerosis and other cardiopathies. Increased ROS production can affect fundamental mechanisms that contribute to atherosclerosis: vascular endothelial cell dysfunction (VEGC), monocyte migration, and LDL oxidation (Deng et al., 2020).

Quercetin, a flavonoid found in peas and other foods, has a wide range of biological and pharmacological effects that correlate with a reduction in the incidence of cardiovascular diseases. Endothelial cells are a semi-permeable barrier between blood and vascular walls capable of secreting vasodilator compounds such as nitric oxide and vasoconstrictor (Li et al., 2019). These cells can destroy endothelial homeostasis in pathological states, such as elevated local blood pressure, hypercholesterolemia, elevated residual triglyceride-rich lipoproteins, and altered vasoactive amines. When this happens, the release of vasomotor factors decreases, and that of vasoconstrictors increases, leading to atherosclerosis. Quercetin can reverse endothelial damage by inhibiting oxidative stress-induced nitrification (Chao, 2010; Li et al., 2019). Another mechanism by which this flavonoid protects the vascular endothelium is by inhibiting lipid peroxidation and protein oxidation, as well as regulating lipid metabolism by reducing low-density lipoproteins (LDL) concentration and increasing high-density lipoproteins (HDL) concentration. A high concentration of LDL may play an important role in the development of cardiovascular disease, as it leads to the formation of atheroma plaques in the arteries that clog the vessels (Xiao et al., 2017; Calabro et al., 2018).

High dietary fiber intake is associated with a reduction in the incidence of cardiovascular disease. Some research has shown that fiber can slow down the process of low-density lipoprotein and glucose absorption, reducing the chances of developing coronary heart disease. In this regard, fiber-rich diets have been shown to lower blood pressure, improve serum lipid levels, and reduce markers of inflammation, preventing the development of coronary heart disease (Papandreou et al., 2019). Sandström et al. (1993) evaluated the effect of pea fiber on cardiovascular health. Fiber intake reduced postprandial triglyceride responses, reducing the risk of atheroma plaque formation.

### 3.4. Antihypertensive Capacity

Hypertension is a disease that occurs when systolic and diastolic blood pressure are elevated. The regulation of blood pressure depends on the renin-angiotensin-aldosterone system. Angiotensin-converting enzyme converts inactive angiotensin I into the potent vasoconstrictor angiotensin II and the vasodilator bradykinin into an inactive peptide that causes an increase in blood pressure. Therefore, many drugs used in hypertension management are inhibitors of this enzyme. Many studies have demonstrated the potential of peptides obtained from the hydrolysis of pea protein to inhibit angiotensin-converting enzyme (Stilling, 2020).

Jakubczyk and Baraniak (2014) obtained bioactive peptides from pea protein hydrolysis under simulated gastrointestinal conditions, evaluating their inhibitory activity against the angiotensin-converting enzyme. Thirteen peptide fractions were obtained, but only four showed inhibitory activity. The lowest inhibitory value ($IC_{50}$) was 0.073 mg/mL, corresponding to the fraction with the highest peptide concentration. The amino acid sequence of this fraction showed that hydrophilic amino acids were the main constituents, with glutamine, glutamic acid, alanine, glycine, serine, asparagine, aspartic acid, and arginine. The amino acid sequence of the peptide was determined as follows: GGSGNY (553.41 m/z), DLKLP (585.56 m/z), GSSDNR (634.56 m/z), MRDLK (662.59 m/z), and HNTPSR (711.59 m/z), there is evidence to indicate that the inhibitory activity of the peptides depends on the C-terminal structure of the tripeptide. The most potent inhibitors contain hydrophobic amino acid residues (aromatic or with branched side chains), such as those observed in this study containing proline (DLKLP).

In another study by Liao et al. (2019), peptides with angiotensin-converting enzyme regulatory activity were identified. A pea protein hydrolysate was obtained by enzymatic hydrolysis using thermoase and pepsin. Three fractions with inhibitory activity were identified with the following amino acid sequences: AKSLSDRFSY, VPHYNL, and VRASSNLN. These three peptides were reported for the first time. Of the peptides obtained, only AKSLSDRFSY showed a significant effect on enzyme regulation. This peptide results from the hydrolysis of globulins present in peas. The results of this study confirmed the feasibility of identifying angiotensin-converting enzyme regulatory peptides from pea protein.

Barbana and Boye (2010) evaluated the effect of the pea protein hydrolysate obtained by gastrointestinal digestion on the angiotensin-converting inhibitory enzyme activity. The results showed that the obtained hydrolysates inhibited 70% of enzyme activity. The inhibitory concentration (0.7 mg/mL) reducing enzyme activity by 50% ($IC_{50}$) was 159 ± 1 for pea proteins. Subsequently, Girgih et al. (2016) used the same pea protein isolate to evaluate the short-term (24 h) and long-term (5 weeks) angiotensin inhibitory activity in hypertensive rats. Amino acid analysis showed that peptides with angiotensin-converting enzyme and renin inhibitory activity depicted tetrapleptic and hydrophilic sequences of between 2 and 6 amino acids. The $IC_{50}$ values for renin and angiotensin-converting enzyme inhibition were 0.57 and 0.10 mg/mL, respectively. Administration of the pea peptides to the rats reduced their blood pressure by -22 mm Hg after three weeks. Inhibitory activity was reduced in the 4th and 5th week of administration, although the blood pressure values of the peptide-treated rats decreased in the 4th and 5th weeks. We conclude that the antihypertensive potency of PPH-5 may be due to the presence of an easily absorbed substance.

Studies on the blood pressure-lowering properties of peas have shown that some of the peptides obtained from the hydrolysis of the protein of this legume are effective angiotensin-converting enzyme inhibitors. The efficacy of peptides depends on their amino acid sequence and structure. Therefore, the method used to hydrolyze the protein plays an important role in the antihypertensive activity of peas.

### 3.5. Gastrointestinal Function and Weight Management

Some studies have analyzed the effect of pea consumption on gastrointestinal function and some gastrointestinal symptoms. The fiber contained in the shell and in peas may improve gastrointestinal transit. Flogan and Dahl (2010) demonstrated that the pea hull fiber addition in combination with an inulin supplement to a snack given to children with constipation significantly increased stool frequency with no adverse side effects reported. Consumption of peas, unlike other legumes such as chickpeas and lentils, does not cause flatulence, bloating, or intestinal discomfort.

In addition to regulating gastrointestinal function, consuming pea fiber may reduce body weight, specifically body composition and metabolic markers. Increasing fiber intake has been associated with body weight reduction because this compound can retain water in its structure, contributing to a feeling of satiety (Slavin, 2013; Ruiz et al., 2020). Lambert et al. (2017) evaluated the effect of yellow pea fiber intake on body composition and metabolic markers in overweight and obese adults. Results showed that pea fiber consumption resulted in a loss of body weight of $0.87 \pm 0.37$ kg, associated with a reduction in body fat percentage. Individuals given fiber reduced their caloric intake by 16%. Without other lifestyle changes, incorporating 15 g/day of yellow pea fiber may produce small but significant metabolic benefits and aid in obesity management. Pea fiber also contains prebiotic oligosaccharides. It can help regulate the growth of bacteria in the gut.

Intestinal microbiota plays indispensable roles in food digestion, boosting immunity, regulating some metabolic functions, and contributing to detoxification. Gut micro-organisms use compounds such as dietary fiber as a source of energy. This process generates secondary metabolites with biological effects. Short-chain fatty acids with neuroactive properties are produced in the gut from pea dietary fiber (Martens et al., 2017). In addition to dietary fiber, pea protein hydrolysates can modulate intestinal bacteria's metabolism and adhesive potential to the cell lining and exert salutary effects on the local environment (Ge et al., 2020). Furthermore, non-α-galacto-oligosaccharides from peas were shown to have prebiotic effects by reducing gas production during colonic fermentation.

Pea shell fiber consumption is also associated with the modulation of gut microbiota, specifically *Bifidobacterium*. Bifidobacteria are beneficial for gut health because they prevent enteric infections and suppress pathogenic bacteria, enhancing the host immune system. Oligosaccharides in pea fiber promote the growth of bifidobacteria in the colon (Ge et al., 2020).

### 3.6. Cancer Prevention

Many scientific studies indicate that legume consumption can be considered an alternative chemo-preventive and therapeutic treatment against several types of cancer, especially colon cancer. Pea extracts have been shown to have anticancer activity (Bello et al., 2018). This activity may be due to their isoflavone content: genistein and diadezin. Genistein showed antiproliferative effects on mitogen-stimulated cell growth of human breast cancer cells. The dietary fiber content may also contribute to reducing the effects of certain types of cancer, especially colorectal and rectal cancer. Several bioactive micronutrients and phytochemicals in peas contribute to a healthy diet and reduce the risk of cancer disorders (Ruiz et al., 2020).

Phenolic compounds present in pea seed hulls also have anticarcinogenic activity. The phenolic profile of seed coat extracts from 10 European pea varieties was evaluated. Extracts of dark-colored varieties showed a high phenolic content (up to 46.56 mg GAE/g), as well as a strong antioxidant activity that could be attributed to the presence of gallic acid, epigallocatechin, naringenin, and apigenin. The obtained phenolic extracts showed concentration-dependent cytotoxic effects on LS174 human colon adenocarcinoma malignant cell lines, MDA-MB-453 human breast carcinoma, A594 human lung carcinoma, and K562 myelogenous leukemia. The cytotoxic activity of the extracts is closely related to the epigallocatechin and luteolin contents. Cell cycle analysis in LS174 cells in the presence of a caspase-3 inhibitor indicates that the extracts can activate other modes of cell death besides caspase-3-dependent apoptosis. Therefore, extracts from the seed hulls of dark-colored pea varieties could be used as potential cancer chemo-preventive agents (Stanisavljevic et al., 2016).

## Conclusion

The effects of consuming peas or some of their compounds on human health have been evaluated. These effects are mainly due to their antioxidant compounds and protein. Antioxidant compounds cause the activation of antioxidant enzymes and reduction of lipid peroxidation, amending diabetes-related health complications. Other mechanisms produced by pea consumption were an insulin release increase through pancreatic β-cell regeneration and activation of insulin-induced lipoprotein hydrolysis. Pea protein, especially its hydrolysates, has antioxidant, antihypertensive, and gut bacteria-modulating properties, which help to control the symptoms of some cardiovascular and intestinal diseases. However, further research on pea protein and its bioactive peptides is necessary to fully elucidate their mechanisms of action and their potential to enhance human health.

## References

American Diabetes Association (2017) 4. Lifestyle management. Diabetes Care 40(S1):S33–S43.

Adwas, A. A., Elsayed, A. S. I., Azab, A. E. & Quwaydir, F. A. (2019). Oxidative stress and antioxidant mechanisms in human body. *Journal of Applied Biotechnology & Bioengineering, 6*, 43–47. doi:10.15406/jabb.2019.06.00173.

Barbana, C. & Boye, J. I. (2010). Angiotensin I-converting enzyme inhibitory activity of chickpea and pea protein hydrolysates. *Food Research International*, 43 (6), 1642–1649. doi:10.1016/j.foodres.2010.05.003.

Bello, O. A., Ayanda, O. I., Aworunse, O. S., Olukanmi, B. I., Soladoye, M. O., Esan, E. B., & Obembe, O. O. (2018). Solaneciobiafrae: An underutilized nutraceutically-important African indigenous vegetable. *Pharmacognosy Reviews*, 12(23), 128–132. doi:10.4103/phrev.phrev_43_17.

Borges-Martínez, E., Gallardo-Velázquez, T., Cardador-Martínez, A., Moguel-Concha, D., Osorio-Revilla, G., Ruiz-Ruiz, J. C. & Martínez, C. J. (2021). Phenolic compounds profile and antioxidant activity of pea (*Pisum sativum* L.) and black bean (*Phaseolus vulgaris* L.) sprouts. *Journal of Food Science and Technology*, 2061, 1–7. doi:10.1590/fst.45920.

Brennan, C. S. (2005). Dietary fibre, glycaemic response, and diabetes. *Molecular Nutrition and Food Research*, 49 (6), 560–570. doi:10.1002/mnfr.200500025.

Burger, T. G., & Zhang, Y. (2019). Recent progress in the utilization of pea protein as an emulsifier for food applications. *Trends in Food Science & Technology*, 86, 25–33. doi:10.1016/J.TIFS.2019.02.007.

Cade, W. T. (2008). Diabetes-related microvascular and macrovascular diseases in the physical therapy setting. *Physical Therapy*, 88, 1322–1335. doi:10.2522/ptj.20080008.

Calabro, V., Litterio, M. C., Fraga, C. G., Galleano, M. & Piotrkowski, B. (2018). Effects of quercetin on heart nitric oxide metabolism in l-NAME treated rats. *Archives of Biochemistry and Biophysics*, 647, 47–53. doi:10.1016/j.abb.2018.03.041.

Chao, Y. (2010). Effects of quercetin on vascular endothelial cell growth and NO production. *Journal of Dalian University*, 31 (6), 84–86.

Costantini, M., Summo, C., Centrone, M., Rybicka, I., D'Agostino, M., Annicchiarico, P., Caponio, F., Pavan, S., Tamma, G. & Pasqualone, A. (2021). Macro– and micro–nutrient composition and antioxidant activity of chickpea and pea accessions. *Polish Journal of Food and Nutrition Sciences*, 71 (2), 177–185.doi: 10.31883/pjfns/135813.

Dahl, W. J., Foster, L. M. & Tyler, R. T. (2012). Review of the health benefits of peas (*Pisum sativum* L.). *British Journal of Nutrition* 108 (S1), S3–S10. doi:10.1017/S0007114512000852.

Daryanto, S., Wang, L. & Jacinthe, P. A. (2015). Global synthesis of drought effects on root and tuber production. *PLOS ONE*, 10(6), e0127401. doi:10.1371/journal.pone.012740.

De Angelis, D., Pasqualone, A., Allegretta, I., Porfido, C., Terzano, R., Squeo, G., & Summo, C. (2021). Antinutritional factors, mineral composition and functional properties of dry fractionated flours as influenced by the type of pulse. *Heliyon*, **7**(2), e06177. doi:10.1016/j.heliyon.2021.e06177.

Deng, Q., Li, X. X., Fang, Y., Chen, X., & Xue, J. (2020). Therapeutic potential of quercetin as an antiatherosclerotic agent in atherosclerotic cardiovascular disease: A review. *Evidence-based Complementary and Alternative Medicine*, 5926381. doi:10.1155/2020/592638.

Djoullah, A., Husson, F., & Saurel, R. (2018). Gelation behaviors of denaturated pea albumin and globulin fractions during transglutaminase treatment. *Food Hydrocolloids*, *77*, 636-645. doi:10.1016/j.foodhyd.2017.11.005.

Dueñas, M., Hernández, T. & Estrella, I. (2006). Assessment of in vitro antioxidant capacity of the seed coat and the cotyledon of legumes in relation to their phenolic contents. *Food Chemistry*, 98, 95–103. doi:10.1016/j.foodchem.2005.05.052.

Fahim, J. R., Attia, E. Z., & Kamel, M. S. (2019). The phenolic profile of pea (*Pisum sativum*): A phytochemical and pharmacological overview. *Phytochemistry Reviews*, 18(1), 173–198. doi:10.1007/s11101-018-9586-9.

FAOSTAT. (2019). http://www.fao.org/faostat/en/#data/QC.

Flogan, C., & Dahl, W. (2010). Effects of fiber-fortified foods on children with constipation: potential improved stool frequency and decreased energy intake. *Infant, Child, & Adolescent Nutrition*, 2(5), 312–317. doi:10.1177/1941406410383980.

Fox SI. (2017). Human Physiology. 16th Ed. McGraw-Hill Education, 666–670. ISBN10: 1260720462. https://www.mheducation.com/highered/product/human-physiology-fox-rompolski/M9781260720464.html.

Ge, J., Sun, C. X., Corke, H., Gul, K., Gan, R. Y., & Fang, Y. (2020). The health benefits, functional properties, modifications, and applications of pea (*Pisum sativum* L.) protein: Current status, challenges, and perspectives. *Comprehensive Reviews in Food Science and Food Safety*,19, 1835–1876. doi:10.1111/1541-4337.12573.

Girgih, A. T., Chao, D. F., Lin, L., He, R., Jung, S., &Aluko, R. E. (2015). Enzymatic protein hydrolysates from high pressure-pretreated isolated pea proteins have better antioxidant properties than similar hydrolysates produced from heat pretreatment. *Food Chemistry*, 188, 510–516. doi:10.1016/j.foodchem.2015.05.024.

Girgih, A. T., Nwachukwu, I. D., Onuh, J. O., Malomo, S. A., & Aluko, R. E. (2016). Antihypertensive Properties of a Pea Protein Hydrolysate during Short- and Long-Term Oral Administration to Spontaneously Hypertensive Rats. *Journal of Food Science*, 81(5), H1281–H1287. doi:10.1111/1750-3841.13272.

Guo, F., Xiong, H., Wang, X., Jiang, L., Yu, N., Hu, Z., & Tsao, R. (2019). Phenolics of green pea (*Pisum sativum* L.) hulls, their plasma and urinary metabolites, bioavailability, and in vivo antioxidant activities in a rat model. *Journal of Agricultural and Food Chemistry*, 67(43), 11955–11968. doi:10.1021/acs.jafc.9b04501.

Halliwell B. (2007). Biochemistry of oxidative stress. *Biochemical Society Transactions*, 35, 1147–1150. https://citeseerx.ist.psu.edu/document?repid=rep1&type=pdf&doi=a638fd8bd6d3391ef700d2d6725d74d2a14ee969.

Holman, R. R., Paul, S. K., Bethel, M. A., Matthews, D. R. & Neil, H. (2008). 10-year follow-up of intensive glucose control in type 2 diabetes. *The New England Journal of Medicine*, 359, 1577–1589. doi:10.1056/NEJMoa0806470.

Ito, F., Sono, Y. & Ito, T. (2019). Measurement and clinical significance of lipid peroxidation as a biomarker of oxidative stress: Oxidative stress in diabetes, atherosclerosis, and chronic inflammation. *Antioxidants*, 8, 72-100. doi:10.3390/antiox8030072.

Jakubczyk, A., & Baraniak, B. (2014). Angiotensin I converting enzyme inhibitory peptides obtained after in vitro hydrolysis of pea (*Pisum sativum* var. Bajka) globulins. *BioMed Research International*, 438459. doi:10.1155/2014/438459.

Jideani, A. I. O., Silungwe, H., Takalani, T., Omolola, A. O., Udeh, H. O. & Anyashi, T. A. (2021). Antioxidant-rich natural fruit and vegetable products and human health. *International Journal of Food Properties*, 24, 41–67. doi:10.1080/10942912.2020.1866597.

Khan, T. N., Ramzan, A., Jillani, G. & Mehmood, T. (2013). Morphological Performance of Peas (*pisum sativum*) Genotypes under Rainfed conditions of Potowar Region. *Journal of Agricultural Research,* 51(1), 51- 60. https://apply.jar.punjab.gov.pk/upload/1373835770_72_549__51p6.pdf.

Kumari, T. & Deka, S. C. (2021). Potential health benefits of garden pea seeds and pods: A review. *Legume Science*, 3, e82. doi:10.1002/leg3.82.

Lam, A. C. Y., Can Karaca, A., Tyler, R. T. & Nickerson, M. T. (2018). Pea protein isolates: Structure, extraction, and functionality. *Food Reviews International*, 34, 126–147. doi:10.1080/87559129.2016.1242135.

Lambert, J. E., Parnell, J. A., Han. J., Sturzenegger, T., Paul, H. A., Vogel, H. J. & Reimer, R. A. (2014). Evaluation of yellow pea fibre supplementation on weight loss and the gut microbiota: a randomized controlled trial. *BMC Gastroenterology*, 14-69. https://bmcgastroenterol.biomedcentral.com/articles/10.1186/1471-230X-14-69.

Lambert, J. E., Parnell, J. A., Tunnicliffe, J. M., Han, J., Sturzenegger, T. & Reimer, R. A. (2017). Consuming yellow pea fibre reduces voluntary energy intake and body fat in overweight/obese adults in a 12-week randomized controlled trial. *Clinical Nutrition*, 36(1), 126–33. doi:10.1016/j.clnu.2015.12.016.

Li, M., Wang, L. & Chang, B. (2019). Research progress in the mechanism of vascular endothelial cell dysfunction. *Chinese Journal of Arteriosclerosis*, 27 (8), 730–736.

Li, R. Y., Dai, T. T., Tan, Y. B., Fu, G. M., Wan, Y., Liu, C. M. & McClements, D. J. (2020). Fabrication of pea protein-tannic acid complexes: Impact on formation, stability, and digestion of flaxseed oil emulsions. *Food Chemistry*, 310, 125828. doi:10.1016/j.foodchem.2019.125828.

Liao, W., Fan, H., Liu, P. & Wu, J. (2019). Identification of angiotensin converting enzyme 2 (ACE2) up-regulating peptides from pea protein hydrolysate. *Journal of Functional Foods*, 103395. doi:10.1016/j.jff.2019.05.051.

Lunde, M. S., Hjellset, V. T., Holmboe-Ottesen, G. & Høstmark, A. T. (2011). Variations in postprandial blood glucose responses and satiety after intake of three types of bread. *Journal of Nutrition and Metabolism*, 437587. doi:10.1155/2011/437587.

Martens, L. G., Nilsen, M. M. & Provan, F. (2017) Pea hull fibre: novel and sustainable fibre with important health and functional properties. *EC Nutrition*, 10, 139–148. https://www.am-nutrition.no/wp-content/uploads/2017/11/AMN-Pea-hull-fiber-article-ECNU-10-00345.pdf.

Mejri, F., Khoud, H. B., Njim, L., Baati, T., Selmi, S., Serralheiro, M., Rauter, A. P., Hosni, K. & Hosni, K. (2019). *In vitro* and *in vivo* biological properties of pea pods (*Pisum sativum* L.). *Food Bioscience*, 32, 100482. doi:10.1016/j.fbio.2019.100482.

Mertens, C., Dehon, L., Bourgeois, A., Verhaeghe-Cartrysse, C. & Blecker, C. (2012). Agronomical factors influencing the legumin/vicilin ratio in pea (*Pisum sativum* L.) seeds. *Journal of the Science of Food and Agriculture*, 92, 1591–1596. doi:10.1080/87559129.2016.1242135.

Ozcan, T., Akpinar-Bayizit, A., Yilmaz-Ersan, L. & Delikanli, B. (2014). Phenolics in Human Health. *International Journal of Chemical Engineering and Applications*, 5 (5), 393-396. doi:10.7763/IJCEA.2014.V5.416.

Papandreou, C., Becerra-Tomás, N., Bulló, M., Martínez-González, M. A., Corella, D., Estruch, R., & Serra-Majem, L. (2019). Legume consumption and risk of all-cause, cardiovascular, and cancer mortality in the PREDIMED study. *Clinical Nutrition*, 38(1), 348–356. doi:10.1016/j.clnu.2017.12.019.

Parihar, A. K., Dixit, G. P., Bohra, A., Gupta S. D., & Singh A. K. (2020). Genetic Advancement in Dry Pea (*Pisum sativum* L.): Retrospect and Prospect. In S. S. Gosal,

& S. H. Wani (Eds.), *Accelerated Plant Breeding*, 3 (283-341). Cham: Springer International Publishing. doi:10.1007/978-3-030-47306-8_10.

Pownall, T. L., Udenigwe, C. C., & Aluko, R. E. (2010). Amino acid composition and antioxidant properties of pea seed (*Pisum sativum* L.) enzymatic protein hydrolysate fractions. *Journal of Agricultural and Food Chemistry*, 58(8), 4712–4718. doi:10.1021/jf904456r.

Pownall, T. L., Udenigwe, C. C., & Aluko, R. E. (2011). Effects of cationic property on the in vitro antioxidant activities of pea protein hydrolysate fractions. *Food Research International*, 44, 1069–1074. doi:10.1016/j.foodres.2011.03.017.

Qin, G., Xu, W., Liu, J., Zhao, L. & Chen, G. (2021). Purification, characterization and hypoglycemic activity of glycoproteins obtained from pea (*Pisum sativum* L.). *Food Science and Human Wellness*, 10:297–307. doi:10.1016/j.fshw.2021.02.021.

Ramos, S. (2008). Cancer chemoprevention and chemotherapy: Dietary polyphenols and signaling pathways. *Molecular Nutrition and Food Research*, 52, 507-526. doi:10.1002/mnfr.200700326.

Rubio, L. A., Pérez, A., Ruiz, R., Guzmán, M. Á., Aranda-Olmedo, I., & Clemente, A. (2014). Characterization of pea (*Pisum sativum*) seed protein fractions. *Journal of the Science of Food and Agriculture*, *94*(2), 280-287. doi:10.1002/jsfa.6250.

Rungruangmaitree, R. & Jiraungkoorskul, W. (2017). Pea, *Pisum sativum*, and Its Anticancer Activity. *Pharmacognosy Reviews*, 11, 39-42. doi:10.4103/phrev.phrev_57_16.

Ruiz, R., Olías, R., Clemente, A., & Rubio, L. A. (2020). A pea (Pisum sativum L.) seed vicilins hydrolysate exhibits PPARγ ligand activity and modulates adipocyte differentiation in a 3T3-L1 cell culture model. *Foods*, 9(6), 793. doi:10.3390/foods9060793.

Sampson, M. J., Gopaul, N., Davies, I. R., Hughes, D. A. & Carrier, M. J. (2002). Plasma F2 isoprostanes: Direct evidence of increased free radical damage during acute hyperglycemia in type 2 diabetes. *Diabetes Care*, 25, 537–541. doi:10.2337/diacare.25.3.537.

Sandström, B., Hansen, L. T. & Sørensen, A. (1994) Pea fibre lowers fasting and postprandial blood triglyceride concentrations in humans. *The Journal of Nutrition*, 124, 2386–2396. doi:10.1093/jn/124.12.2386.

Sato, K., Miyasaka, S., Tsuji, A., & Tachi, H. (2018). Isolation and characterization of peptides with dipeptidyl peptidase IV (DPPIV) inhibitory activity from natto using DPPIV from Aspergillus oryzae. *Food Chemistry*, 261, 51. doi:10.1016/j.foodchem.2018.04.029.

Schafer, G., Schenk, U., Ritzel, U., Ranmdori, G., & Leonhardt, U. (2003). Comparison of the effects of dried peas with those of potatoes in mixed meals on postprandial glucose and insulin concentrations in patients with type 2 diabetes. *American Journal of Clinical Nutrition*, 78, 99–103. doi:10.1093/ajcn/78.1.99.

Slavin, J. (2013). Fibre and prebiotics: Mechanisms and health benefits. Nutrients, 5(4), 1417–1435. doi:10.3390/nu5041417.

Stanisavljević, N. S., Ilić, M. D., Matić, I. Z., Jovanović, Ž. S., Cupić, T., Dabić, D. C., Natić, M. M. & Tešić, Ž. L. (2016). Identification of phenolic compounds from seed coats of differently colored European varieties of pea *(Pisum sativum* L.) and characterization of their antioxidant and In vitro anticancer activities. *Nutrition and Cancer*, 68, 988–1000. doi:10.1080/01635581.2016.1190019.

Stilling, K. (2020). Health Benefits of Pea Protein Isolate: A Comparative Review. *Studies by Undergraduate Researchers at Guelph*, 12, 1–12. doi:10.21083/surg.v12i1.6111.

Tan, S. Y., Siow, P. C., Peh, E., & Henry, C. J. (2018). Influence of rice, pea and oat proteins in attenuating glycaemic response of sugar sweetened beverages. *European Journal of Nutrition*, 57, 2795–2803. doi:10.1007/s00394-017-1547-3.

Tasyurek, M. H., Altunbas, H. A., Canatan, H., Griffith, T. S. & Sanlioglu, S. (2014). GLP-1-mediated gene therapy approaches for diabetes treatment. *Expert Reviews in Molecular Medicine*. 16(e7):1–20. doi:10.1017/erm.2014.7.

Tulbek, M. C., Lam, R. S., Wang, Y. C., Asavajaru, P. & Lam, A. (2017). Pea: a sustainable vegetable protein crop. In: R. N. Sudarshan, J. P. D. Wanasundara, and L. Scanlin, editors. *Sustainable protein sources*. San Diego (CA): Academic Press, 145–164. doi:10.1016/B978-0-12-802778-3.00009-3.

Valko, M., Leibfritz, D. & Moncol, J. (2007). Free radicals and antioxidants in normal physiological functions and human disease. *The International Journal of Biochemistry & Cell Biology*. 2007; 39(1):44–84. doi:10.1016/j.biocel.2006.07.001.

Wang, L., Zhang, X. X., Liu, F. R., Ding, Y. Y., Wang, R., Luo, X. H. & Chen, Z. (2017). Study of the functional properties and anti-oxidant activity of pea protein irradiated by electron beam. Innovative *Food Science and Emerging Technologies*, 41, 124–129. doi:10.1016/j.ifset.2017.01.005.

Weinreb, O., Mandel, S., Amit, T., Moussa, B. & Youdim, H. (2004). Neurological mechanisms of green tea polyphenols in Alzheimer's and Parkinson's diseases. *The Journal of Nutritional Biochemistry*, 15, 506-516. doi:10.1016/j.jnutbio.2004.05.002.

Wei, Y., Zhang, R., Fang, L., Qin, X. Y., Cai, M., & Gu, R. Z. (2019). Hypoglycaemic effects and biochemical mechanisms of Pea oligopeptide on high-fat diet and streptozotocin-induced diabetic mice. *Journal of Food Biochemistry*, 43, Article e13055. doi:10.1111/jfbc.13055.

Xiao, L., Liu, L. & Guo, X. (2017). Quercetin attenuates high fat diet-induced atherosclerosis in apolipoprotein E knockout mice: a critical role of NADPH oxidase. *Food and Chemical Toxicology*, 105, 22–33. doi:10.1016/j.fct.2017.03.048.

Zha, F. C., Dong, S. Y., Rao, J. J., & Chen, B. C. (2019). Pea protein isolate-gum Arabic Maillard conjugates improves physical and oxidative stability of oil-in-water emulsions. *Food Chemistry*, 285, 130–138. doi:10.1016/j.foodchem.2019.01.151.

Zilani, M. N., Sultana, T., Rahman, S. A., Anisuzzman, M., Islam, M. A., Shilpi, J. A. & Hossain, M. G. (2017). Chemical composition and pharmacological activities of *Pisum sativum*. BMC Complementary Medicine and Therapies, 17, 1716. doi:10.1186/s12906-017-1699-y.

# Chapter 12

# Biological Activity of *Phaseolus coccineus, Lupinus, Lens culinaris,* and *Vicia faba*

**E. B. León-Espinosa[1]**
**M. Díaz-Ramírez[2]**
**L. Wallander-Compeán[3]**
**E. Correa-Solís[4]**
**D. Girón-Orozco[5]**
**and M. Rentería-Ortega[1,*]**
[1]Tecnológico Nacional de México, Tecnológico de Estudios Superiores de San Felipe del Progreso, San Felipe del Progreso, México
[2]Departamento de Ciencias de la Alimentación, Universidad Autónoma Metropolitana-Unidad Lerma, México
[3]Instituto Politécnico Nacional CIIDIR Unidad Durango, Durango, México
[4]Universidad de la Cañada. Licenciatura en Nutrición, Teotitlán de Flores Magón, Oaxaca, México
[5]Facultad de Ciencias Agrícolas de la Universidad Autónoma del Estado de México, Licenciatura en Ingeniero Agrónomo Industrial, Toluca, México

## Abstract

Legumes are widely recognized and consumed globally and are second only to cereals in terms of their importance as a source of sustenance. They are highly valued for their biological activity and are considered a sustainable alternative due to their cost-effectiveness. Legumes belong to the *Leguminoceae* family and include various types, such as broad beans, runner beans, lentils, and mung beans. These legumes contain bioactive

---

* Corresponding Author's Email: minervarenteria22@gmail.com

In: Legumes: Beyond Their Nutritional Value
Editors: C. Jiménez-Martínez, M. F. Juárez-Chairez and M. S. Cid-Gallegos
ISBN: 979-8-89530-440-2

compounds like phytates, saponins, and polyphenols/tannins, which have ambivalent nutritional properties depending on their dietary concentration. Understanding these bioactive components and their mechanisms of action in human health is still a challenge in some cases. Research has shown that most bioactive compounds in legumes have antioxidant activities, crucial in preventing various types of cancer and mitigating the risks associated with pathologies like diabetes, cardiovascular diseases, and chronic degenerative disorders. Additionally, legumes have been correlated with favorable health effects, including hypocholesterolemic, antiatherogenic, and anti-inflammatory activities. This chapter will examine the mechanisms underlying the biological activities attributed to the bioactive compounds found in legumes, focusing on examples like runner beans, lupins, lentils, and broad beans.

**Keywords**: legumes, bioactive compounds, antioxidant, anticancer

## 1. Introduction

The Fabaceae plant family is known for its diverse range of seeds in different shapes, sizes, and colors. These seeds contain a variety of bioactive compounds. Legumes, in particular, are renowned for their high protein and complex carbohydrate content, including dietary fiber and resistant starch. They are also low in fat and provide essential minerals such as iron (Fe), zinc (Zn), and calcium (Ca) (Looi et al., 2023). Legumes are an excellent source of proteins, peptides, and phytochemicals, containing substantial amounts of these bioactive components. They also provide bioactive compounds that help prevent chronic diseases. Many studies have shown that consuming legumes has numerous health benefits, mainly due to their high fiber content, both soluble and insoluble, and starch resistance to digestive degradation. Moreover, specific proteins in legumes possess bioactive properties, and polyphenolic compounds further enhance their health-promoting features. However, it's important to note that legumes also have compounds like tannins, saponins, phytates, lectins, and protease inhibitors that may benefit health (Kumar et al., 2021). These compounds may also be considered anti-nutritive factors as they can reduce nutrient absorption, mineral availability, and protein digestibility.

This chapter presents scientific evidence demonstrating the health benefits of consuming legumes, specifically lupins, lentils, broad beans, and *Phaseolus coccineus*. Incorporating these legumes into the diet may aid in the prevention of type 2 diabetes, hypercholesterolemia, hypertension, cancer, and inflammatory conditions. (Arnoldi et al., 2015; Serrano et al., 2016) Epidemiological studies, animal experiments, and human intervention studies have all demonstrated the benefits of consuming legumes. These benefits include enhanced cardiovascular health, weight management, improved metabolism, gastrointestinal health, and effective management of type-2 diabetes (Singh et al., 2017). The effects on the body can differ significantly after consuming different types of legumes, even within the same plant species. These differences are due to variations in the composition of polysaccharides, the amount and type of dietary fiber and starch, protein composition, and the type and quantity of specific phytochemical compounds. Further research is required to comprehend how these phytochemical compounds interact with one another and impact the body in daily dietary scenarios.

## 2. Biological Activities of Legumes

A body of research has recently revealed the beneficial biological activities found in legumes, some of which are listed in Table 1. These encompass antioxidant, immune-regulatory, gastrointestinal-protective, anti-tumor, and anti-bacterial effects. Moreover, their potential extends to the prevention of hypertension, diabetes, and heart disease (Li et al., 2022; Jia et al., 2015).

The significance of polysaccharides has been prominently underscored, with numerous investigations emphasizing their pivotal role as biologically active components within legumes. It has been elucidated that the structural intricacies of polysaccharides are intimately linked to their biological activity, as expounded upon by Liu et al. (2021).

### 2.1. Phytochemical Compounds of Legumes

Legumes are rich in bioactive compounds, although they are present in small amounts. These compounds are known as non-nutritional, anti-nutritional, or phytochemical compounds, and they are an essential part of the plant's defense system against pests and environmental stressors. There are two types of these compounds: naturally thermostable and thermolabile. Heat-stable factors include tannins and phytic acid, while heat-labile factors include protein-based protease inhibitors and lectins (Ha & Luan, 2022).

**Table 1.** Biological activities from different legumes

| Species (common name) | Biological activity | Reference |
|---|---|---|
| *Lens culinaris* | Antioxidant<br>Anti-fungal<br>Anti-bacterial<br>Anti-cancer<br>Anti-diabetic<br>Anti-satiatic | Kripi et al., 2019 |
| *Vicia faba* | Antioxidant<br>Anti-diabetic<br>Enzyme-Inhibition<br>Anti-microbial<br>Anti-Parkinson<br>Anti-convulsant<br>Diuretic<br>Anti-coagulant<br>Anti-cancer<br>Anti-Inflammatory<br>Anti-depressant<br>Anti-hepatotoxic | Alam and Najam, 2022 |
| *Lupinus* | Anti-inflammatory<br>Anti-osteoclastogenic<br>Neuroprotective<br>Analgesic<br>ACE Inhibitor | Lemus et al., 2023,<br>Montserrat et al., 2019<br>Santos et al., 2022<br>Muñoz et al., 2018 |
| *Phaseolus coccineus* | Anti-diabetic<br>Anti-cancer<br>Antioxidant<br>Hypocholesterolemic | Lopez, 2020 |
| *Inga* | Anti-fungal<br>Inhibitory | Lima et al., 2018 |

## 2.2. Antioxidant Activity of Phytochemical Compounds

The beneficial role of legumes for human health also resides in their bioactive molecules of which phenolic compounds, saponins, peptides, and small proteins are the most relevant. Legumes, including lupins, lens, broad beans, and Phaseolus coccineous, are some of the most important crops in the world because of their nutritional quality. These legumes have in common certain primary compounds such as proteins, carbohydrates, lipids, and fiber, as well as secondary products such as phytochemicals (phenolic compounds, flavonoids, tannins, polyphenols) that are considered bioactive compounds, however, each of them varies in quantity and composition.

It was known that phytochemical compounds in foods can generate various health benefits, including the prevention of various chronic degenerative diseases associated with oxidative stress such as diabetes, cancer, cardiovascular diseases, and Alzheimer's, among others (Karamać et al., 2018; Amarowicz et al., 2009). Phenolic compounds exert a wide range of beneficial properties, including metabolism, homeostasis, and cell proliferation modulation. Moreover, they have antioxidant properties, which are essential for the prevention of oxidative stress conditions and diseases.

Tannins are another class of polyphenols that are common in the seed coat of many legumes such as beans (Jin et al., 2012). Often, these phytochemicals are present as condensed tannins, which are considered anti-nutritional compounds since they affect protein digestibility, decreasing the absorption of minerals and some vitamins (Jin et al., 2012). However, in the last years, several health benefits have been attributed to the intake of tannins, ranging from antioxidants to the inhibition of lipid peroxidation and antimicrobial activities (Silva & Pogaˇcnik, 2020). Another promising bioactive compound known as Bowman-Birk inhibitor (BBI) is a promising anticancer agent, that showed a proteasome 20 S inhibition action in breast cancer cells.

Bioactive compounds that possess antioxidant properties can slow down the process of oxidation of important biomolecules. It has been observed that the overproduction of free radicals plays a significant role in the development of chronic diseases such as Alzheimer's, various types of cancer, and diabetes (Cerqueni et al., 2021). Numerous studies have identified antioxidant activities within various legume species, establishing a positive correlation between these activities and the total phenolic content. The efficacy of antioxidants is intricately linked to the chemical composition of phenolic compounds, as elucidated by Djordjevic et al. (2011) and Kervin (2004).

On the other hand, lupinus seeds contain various bioactive compounds such as alkaloids, polyphenols, carotenoids, tocopherols, and phytosterols, which have demonstrated antimicrobial, antioxidant, anti-inflammatory, and anti-cancer activities. Mohamed et al., (2017) demonstrated that lupinus has different phytochemical constituents such as alkaloids, anthocyanins, flavonoids, polyphenols, proteins, and glycosides. These constituents extracted in the alcoholic solution presented a DPPH radical trapping activity of 78.73%. The free radical scavenging activities of extract depend on the ability of antioxidant compounds to lose hydrogen and the structural conformation of these components (Shimada et al., 1992).

*V. faba* beans composition (carbohydrates, fibers, proteins, vitamins, and minerals) and non-protein antinutritional factors (saponins, phytic acid, alkaloids, and tannins). Recent papers suggest that *Faba bean* and derivatives could represent a suitable food in the treatment of diabetics, in hypertension. The molecular mechanisms, and antioxidant, anti-inflammatory, and anti-diabetic properties by which faba bean could be involved in the protection against the development of human disease.

**Table 2.** Compounds of some legumes and their antioxidant activity

| Legumes | Bioactive compounds | Assay antioxidant | References |
|---|---|---|---|
| *Vicia faba* | Phenolic<br>Flavonoids | FRAP<br>1.157 mol/kg<br>Oxigen radical absorbance capacity ORAC<br>109 mmol/kg to 149 mmol/kg Trolox equivalents | Turco et al., 2016 |
| | Total phenolics (21.5 mg GAE/g)<br>Condensed tannins (47.7 mg CE/g) | DPPH<br>Antioxidant activity 22.9 mg TROLOX/g | Karataş et al., 2017 |
| *Lens culinaris* | Total polyphenols expressed in caffeic acid 358.26 mg<br>Flavonoids expressed in quercetol<br>1790 mg | DPPH<br>Expressed $IC_{50}$<br>89.27% | Butu et al., 2014 |
| | Polyphenols | ABTS radical Scavenging activity (1.011 mM Trolox eq/mg)<br>FRAP<br>(1.539 Mm eq/mg) | Kang, 2015 |
| *Lupinus* | Total phenolics<br>212 mg GAE/100g<br>Total flavonoids<br>1100 μg catechin/g | DPPH<br>(3.51 mM Trolox)<br>TRAP<br>(0.33 mM Trolox) | Ruíz et al., 2019 |
| | Protein hydrolyzates | ABTS<br>antioxidant activity $IC_{50}$<br>2.0–1.95 μmol/mg | Chirinos et al., 2020 |
| | Phenolic compounds<br>21.2 to 37.1 mg GAE/g | DPPH<br>$IC_{50}$ = 0.32- 1.29 mg/ml | Karamac et al., 2018 |

2, 2, Diphenil -1-picrylhydrazyl (DPPH), 2, 2,-azinobis-(3ethylbenzothiazilodine-6 sulfonic acid (ABTS), Ferric Reducing antioxidant power assay (FRAP), Oxygen radical absorbance capacity (ORAC), Total peroxil radical trapping potential (TRAP), mg gallic acid equivalents (GAE).

Lentils contain bioactive phytochemicals such as phenolics, carotenoids, tocopherols, saponins, phytic acid, and phytosterols. Consumption of lentils has been associated with the risk reduction of many diseases due to the potential antioxidant activity and anti-inflammatory potential of phytochemicals in lentils (Zhang et al., 2015).

Table 2 shows the compounds of some legumes to which antioxidant activity is attributed, as well as the method by which it was determined.

## 2.3. Cardiovascular Protection

Cardiovascular diseases (CVDs), such as ischemic heart disease (IHD) and stroke, are the primary causes of global mortality and disability. Common risk factors for CVDs include dyslipidemia, hypertension, and type 2 diabetes. Studies have shown that consuming legume seeds can help protect against cardiovascular diseases by reducing blood lipid levels and blood pressure (Ros, 2013).

Legumes are an excellent source of phytosterols that can help lower cholesterol absorption in the intestine, thereby reducing the risk of coronary heart disease (CHD) (Li et al., 2022; Marventano et al., 2017). Notably, β-sterols, found abundantly in legumes, contribute to lowering serum cholesterol, blood pressure, and the overall risk of coronary heart disease (Makhmudova et al., 2021).

In general, legumes exhibit cardiovascular protective effects by mitigating hypertension and improving dyslipidemia, evidenced by enhancements in high-density lipoprotein cholesterol (HDLC), reductions in low-density lipoprotein cholesterol (LDL-C), triglycerides (TAG), and blood pressure.

## 2.4. Other Bioactivities of Legumes

Legumes have been found to possess anti-obesity and anti-diabetic properties in addition to their other benefits. They help regulate blood glucose levels and play a protective role against diabetes by containing resistant starch (Non-Starch Polysaccharides or NSPs) (Moreno et al., 2020; Singhal et al., 2014).

## 3. Runner Bean (*Phaseolus coccineus*)

### 3.1. Biological Activity

Runner beans are highly nutritious and contain essential components such as protein, carbohydrates, and minerals. In addition to these, this legume is also rich in phenolic compounds. While it contains antinutrients such as lectin inhibitors, proteases, and saponins. recent studies (Lopez, 2020) suggest that consuming them in low concentrations may be beneficial for human health. Thus, the relationship between the components in runner beans and their impact on human well-being is complex and requires further research.

### 3.2. Bioactivities

The evaluation of antioxidant activity in *P. coccineus* extracts can vary depending on the type of assay used, the extraction methods employed, and the nature of the food matrix. Several in vitro methods have been used to assess the antioxidant activity of *P. coccineus*. Alvarado et al. (2019) have shed light on the antioxidant activity of phenolic compounds found in this plant.

Pan et al. (2015) isolated a dimeric lectin (CHL) from *P. coccineus* var. albonanus Bailey. Their study revealed that CHL exhibited potent antioxidant properties, effectively preventing erythrocyte hemolysis. Furthermore, it demonstrated notable antitumor activity against human nasopharyngeal carcinoma, hepatoma, and breast cancer cells, albeit without antifungal activity. Interestingly, CHL induced a significant dose-dependent increase in the production of pro-inflammatory cytokine mRNAs, suggesting potential therapeutic utility. These findings highlight the multifaceted bioactive properties of *P. coccineus*, encompassing both antioxidant and anti-tumor activities.

Several studies have explored the potential anticancer activity of bioactive compounds found in runner beans. Pan et al. (2015) researched the anti-tumor activity of the dimeric lectin, utilizing the MMT assay. Their findings revealed a significant reduction in total tumor multiplicity in rats fed with purple ayocote bean diets (1.1 times) compared to those on the control diet (3.0 times). This suggests a potential anti-cancer effect associated with the consumption of purple ayocote beans, particularly attributed to the activity of the dimeric lectin.

Studies conducted on *P. coccineus* suggest its potential to ameliorate metabolic disorders and related conditions, including obesity, insulin resistance, dyslipidemia, high blood pressure, and diabetes (Cornier et al., 2008). Another avenue of research highlights the coat of scarlet runner beans, which is rich in proanthocyanidins. These compounds exhibit the ability to inhibit pancreatic lipase activity *in vitro*, signifying their potential as effective anti-obesity agents *in vivo* (Ci et al. 2018). This dual perspective underscores the diverse health-promoting properties of *P. coccineus*, ranging from metabolic disorder mitigation to anti-obesity effects attributed to specific bioactive constituents.

## 4. *Lupinus*

### 4.1. Biological Activity of *Lupinus*

Lupinus is a genus of Fabaceae family plants that have a bushy growth pattern and vibrant inflorescences above the leaves. It is cultivated globally with significant production in Australia and Europe (Cruz et al., 2023). Lupinus can be categorized into two groups: the "Old World" group, comprising 12 annual species primarily found in the Mediterranean, North, and East Africa, and the "New World" group, encompassing numerous annuals, herbaceous perennials, and some shrubs in North and South America.

The nutritional composition of Lupinus varies between the groups. "Old World" lupines, including *L. angustifolius*, *L. albus*, *L. luteus*, *L. hispanicus*, and *L. Micranthus* (Eastwood et al., 2008), are smooth-seeded and have low carbohydrate content (1–4%), high dietary fiber (17%), and a low glycemic index. Lupinus seeds exhibit a mineral content of 3.5%, mainly comprising phosphorus, calcium, magnesium, and potassium (Serrano-Sandoval et al., 2023).

Regarding bioactive compounds, Lupinus contains polyphenols, phytosterols, and peptides, with flavonoids predominating over phenolic acids. Commercial varieties of *L. angustifolius* in Western Australia have reported polyphenol content between 1.6 and 1.9 mg gallic acid equivalents (GAE)/g dry basis and antioxidant capabilities between 2.6 and 5.4 μmol of Trolox equivalents (TE)/g dry basis (Villarino, 2014).

Lupinus has gained attention in various sectors, including food and pharmaceuticals, owing to the rising consumer demand for "naturally functional" products (Liu et al., 2023). Human clinical studies indicate that lupine seed fractions may benefit cardiovascular disease risk factors and metabolic syndrome, with reported benefits on satiety (Kouris-Blazos and Belski, 2016; Villarino et al., 2016). Studies have shown increased satiety after consuming lupine bread compared to wheat-only bread (Pereira et al., 2022; Keogh et al., 2011).

In randomized studies, Lupinus has shown potential in reducing atherosclerotic lesions, as well as lowering total cholesterol and LDL levels. The activity of γ-conglutin in reducing hyperglycemia has shown significant effects comparable to standard antidiabetic medication (Sedláková et al., 2016). Lupine seed fiber has also been identified as a prebiotic ingredient, containing around 40% oligosaccharides, particularly α-galactosides, with the potential to enhance the *Bifidobacterium* population in the colon, supporting a healthy gut (Van de Noort, 2016).

## 5. Biological Activity of Lentil (*Lens culinaris*) and Broad Bean (*Vicia faba L.*)

Polyphenols, which exhibit antioxidant biological activity, are present in *Lens culinaris* and *Vicia faba*. These compounds play a vital role in reducing oxygen concentration, intercepting oxygen, preventing the initiation of the oxidation chain, eliminating initial radicals like hydroxyl radicals, and converting primary oxidation products into non-radical species (Tanveer et al., 2022). Additionally, polyphenols can chelate pro-oxidant metal ions, such as ferrous ions. *Lens culinaris* contains various phenolic compounds, including flavan-3-ols, proanthocyanidins, anthocyanidins, flavonols, stilbenes, flavones, and flavanones, contributing to its antioxidative effects (Alshikh, 2015).

Broad bean (*Vicia faba* L.) is another legume of interest due to its soluble sugars, fatty acids, minerals, proteins, and dietary fiber composition. It also contains flavonoids and polyphenols, providing antioxidant properties and contributing to the prevention of chronic degenerative diseases such as cancer, Parkinson's disease, allergies, diarrhea, and gastric ulcers (Ceramella et al., 2022; Mendez et al., 2022).

In the context of lentils, Wang et al. (2021) identified the biological activity of lentil lectin derived from *Lens culinaris*. This phytochemical inhibited SARS-CoV-2 infection, exhibiting broad anti-SARS-CoV-2 activity against various mutant strains and epidemic variants. Lentil lectin also demonstrated antiviral activity against other coronaviruses such as SARS-CoV and MERS-CoV. Notably, deleting glycosylation sites in the SARS-CoV-2 S protein did not influence lentil lectin susceptibility, suggesting its ability to bind to glycans at multiple sites on the S trimer (Hernández et al., 2020).

Similarly, linolenic acid found in *Vicia faba* (46.1%) has been associated with a significant reduction in the replication of coronavirus 229-E by destabilizing the lipid bilayer through its incorporation into the virus's lipid membrane (Khalil, 2020). This underscores the potential antiviral properties of specific components within *Vicia faba* against coronaviruses.

## 5.1. Anticancer Activity of Lentil (*Lens culinaris*) and Broad Bean (*Vicia faba L.*)

The reported anti-cancer activity of lentils is associated with a reduced incidence of colon, thyroid, liver, breast, and prostate cancers. The polyphenols found in lentils have the potential to bind to cancer cell membranes and receptors, inducing cytotoxicity, apoptosis, and autophagy, ultimately inhibiting tumor growth (Ganesan and Xu, 2017; Caccialupi et al., 2010).

The anti-cancer mechanism of lentil lectins and phenolics involves binding to ribosomes, inhibiting protein synthesis. This process triggers non-apoptotic G1 phase accumulation mechanisms, G2/M phase cell cycle arrest, and apoptosis. Additionally, it activates the caspase cascade in mitochondria, downregulates telomerase activity, and inhibits angiogenesis (Ganesan and Xu, 2017).

Furthermore, *Vicia faba* agglutinin (VFA), a lectin present in broad beans, has been reported by González, (2005) to stimulate morphological differentiation and the malignant phenotype of colon cancer cells. The aggregation of cancer cells induced by VFA can be blocked by a monoclonal antibody against the epithelial cell adhesion molecule (MACep) through binding with N-glycosylated receptors.

González, (2005), describes various mechanisms of lectin-induced anti-cancer action:

a) Binding to the membranes of tumor cells.
b) Cytotoxic effects of lectins on tumor cells (decreased protein synthesis and induction of apoptosis).
c) Reduction of cell proliferation.
d) Stimulation of the immune system (cytokines, interleukin-1 and 6, tumor necrosis factor-alpha [TNF-α], induction of cells to repair damaged DNA).

Lectins engage with the external transmembrane saccharide glycophorin, which governs the membrane of internal cytoskeletons. The lectin's specificity for saccharides determines its binding to glycophorin receptors (González, 2005). Regarding *Vicia faba*, Martineau (2022) notes that digested protein from trypsin containing this legume has demonstrated a protective effect against colon cancer in mice. This effect manifests by reducing the number of pre-neoplastic lesions induced by a high-fat diet or azoxymethane injections. Additionally, broad bean hydrolyzate has exhibited a cholesterol-lowering effect, notably improving the lipid profile, specifically in the biosynthesis of endogenous cholesterol (León *et al*; 2016).

Different varieties of *Vicia faba*, such as Nuria and Rossa, have been shown to impact the proliferation of various cell types, including AGS, HT-29, BL13, and Hep G2. Extracts from the Nuria variety suppress the proliferation of non-transformed CCD-18Co colon cells. In contrast, the Rossa variety influences the proliferation of HT-29 colon cancer cells, promoting apoptosis and necrosis (Siah et al., 2012).

## 5.2. Autoimmune Diseases

In the context of autoimmune diseases, including systemic lupus erythematosus, rheumatoid arthritis, type I diabetes, psoriasis, multiple sclerosis, and Hashimoto's thyroiditis, *Vicia faba* plays a role in the aryl hydrocarbon receptor (AhR) pathway. The AhR, located in the human intestine, generates regulatory T cells (Tregs) and is activated by various components such as quercetin, myricetin, kaempferol, chrysin, apigenin, naringenin, genistein, luteolin, phloretin, daidzein, diosmin, coumestrol, resveratrol, catechin, epicatechin, pelargonidin, tryptophan, and L-DOPA (Méndez, 2022).

The emergence of autoimmune diseases, as explained by Méndez (2022), is associated with a reduction in Tregs levels, indicating greater disease activity and a poor prognosis in autoimmune patients. Tregs, a unique subset of CD4+ T cells, are crucial in maintaining immune tolerance, preventing responses against food or commensal components, and avoiding reactivity against self-antigens. Tregs can inhibit the proliferation and activation of effector T cells through cell-to-cell contact or secretion of TGF-β, IL-10, granzyme, and perforin. The dysbiosis in intestinal microbiota is implicated in the decrease in Tregs.

The genetic control of immune response mediated by AhR is explained by its interaction with the activated B-cell enhancer nuclear factor-kappa light chain (NF-κB) transcription factor, coordinating inflammatory and survival responses. AhR dimerizes with NF-κB subunits such as RelA and RelB, leading to their recruitment to consensus DNA sequences and promoting gene activation. AhR/Rel dimers are found in lymphocytes, which bind to DNA response elements of both transcription factors, supporting the activation of target genes in the AhR and NF-κB pathways (Méndez, 2022).

## Conclusion

In summary, legumes, including unconventional varieties like lupinus, broad beans, and runner beans, are nutritional powerhouses offering diverse bioactive compounds and significant health benefits. As the second most important source of nutrition globally, their sustainable and cost-effective nature makes them a key component in promoting balanced and sustainable diets.

The research underscores the positive impact of legume consumption on cardiovascular health, weight control, metabolic syndrome, gastrointestinal health, and type-2 diabetes. Unconventional legumes exhibit notable biological activities such as antioxidant, immune-regulating, and anti-tumor effects. These legumes and their bioactive compounds, like lectins and polyphenols, showcase potential in preventing various diseases, including hypertension, diabetes, and heart disease.

Furthermore, the anti-cancer properties of certain legumes, evidenced by their ability to inhibit tumor growth and induce apoptosis, add another layer to their health-promoting profile. Despite the presence of anti-nutritional factors, such as tannins and phytic acid, thermal treatments, and postharvest processes can mitigate these effects, enhancing the overall nutritional quality of legumes.

Legumes, especially unconventional varieties, present a holistic approach to preventive nutrition, addressing a spectrum of health challenges. Their multifaceted biological activities and nutritional richness position them as valuable allies in fostering well-being and shaping sustainable, health-conscious dietary practices for the future. Harnessing the full potential of these leguminous treasures is key to advancing nutritional strategies that prioritize both human health and environmental sustainability.

## References

Alshikh, N., de Camargo, A. C., & Shahidi, F. (2015). Phenolics of selected lentil cultivars: Antioxidant activities and inhibition of low-density lipoprotein and DNA damage. *Journal of Functional Foods, 18*, 1022-1038. doi:10.1016/j.jff.2015.05.018.

Alam, T., and Najam, L. (2022). Faba-Bean Antioxidant and Bioactive Composition: Biochemistry and Functionality. In Faba Bean: *Chemistry, Properties and* Functionality (pp. 123-162). Springer International Publishing Ed. doi: 10.1007/978-3-031-14587-2_6.

Alvarado, L. A.N, Gómez, O. L.M, Heredia, J.B, Baeza, J. R, Garcia, G. H.S, Lopez, M. L.X (2019). Nutritional and bioactive characteristics of Ayocote bean (*Phaseolus coccineous* L.). CYTA - *Journal of Food* .17:199–206. DOI:20.500.11799/104724.

Amarowicz, R., Karamać, M., Dueñas, M., & Pegg, R. B. (2017). Antioxidant activity and phenolic composition of a red bean (*Phaseolus vulgaris*) extract and its fractions. *Natural Product Communications, 12*(4). doi: 10.1177/1934578X1701200420.

Arnoldi, A., Boschin, G., Zanoni, C., & Lammi, C. (2015). The health benefits of sweet lupin seed flours and isolated proteins. *Journal of Functional Foods, 18*, 550-563. doi: 10.1016/j.jff.2015.08.012.

Butu, M., Rodino, S., Butu, A., & Butnariu, M. (2014). Screening of bioflavonoid and antioxidant activity of lens culinaris medikus. *Digest Journal of Nanomaterials & Biostructures (DJNB), 9*(2).

Caccialupi, P., Ceci, L. R., Siciliano, R. A., Pignone, D., Clemente, A., & Sonnante, G. (2010). Bowman-Birk inhibitors in lentil: Heterologous expression, functional characterisation and anti-proliferative properties in human colon cancer cells. *Food Chemistry*, 120(4), 1058-1066. doi:10.1016/j.foodchem.2009.11.051.

Ceramella, J., La Torre, C., De Luca, M., Iacopetta, D., Fazio, A., Catalano, A., & Rosano, C. (2022). Exploring the anticancer and antioxidant properties of *Vicia faba* L. pods extracts, a promising source of nutraceuticals. *Peer Journal. 10*, e13683. doi: 10.7717/peerj.13683.

Cerqueni, G., Scalzone, A., Licini, C., Gentile, P., Mattioli-Belmonte, M. (2021). Insights into oxidative stress in bone tissue and novel challenges for biomaterials. *Materials Science and Engineering*. 130:112433. doi: https://d10.1016/j.msec.2021.112433.

Chirinos, R., Cerna, E., Pedreschi, R., Calsin, M., Aguilar-Galvez, A., & Campos, D. (2021). Multifunctional in vitro bioactive properties: Antioxidant, antidiabetic, and antihypertensive of protein hydrolyzates from tarwi (Lupinus mutabilis Sweet) obtained by enzymatic biotransformation. *Cereal Chemistry, 98*(2), 423-433. doi:10.1002/cche.10382.

Ci, Z. H., Jiang, C. Y., Feng, S., Wu, S., Cui, Y., Sasaki, Y., & Kojima, M. (2018). Anti-obesity effect of proanthocyanidins from the coat of scarlet runner beans on high-fat diet-fed mice. *Journal of Food and Nutrition Research*, 6(2), 103-109. doi: 10.12691/jfnr-6-2-6.

Cornier, M.A, Dabelea, D., Hernandez, T. L., Lindstrom, R. C., Steig, A. J., Stob, N. R., Van Pelt, R. E., Wang, H., Eckel, R. H. (2008). The metabolic syndrome. *Endocrine Reviews*. 29(7):777–822. doi:10.1210/er.2008-0024.

Cruz, C. I., Santos, S. G., Álvarez, L. A. I., Pedroche, J., Lardone, P. J., Arnoldi, A., & CarrilloV. A. (2023). Pleiotropic biological effects of *Lupinus* spp. protein hydrolysates. *Trends in Food Science & Technology*.133: 244-266. doi: https://doi.org/10.1016/j.tifs.2023.02.011.

Djordjevic, T. M., Šiler-Marinkovic, S. S., & Dimitrijevic-Brankovic, S. I. (2011). Antioxidant activity and total phenolic content in some cereals and legumes. *International Journal of Food Properties,* 14(1), 175-184. doi:10.1080/10942910903160364.

Eastwood, R. J., Drummond, C. S., Schifino, W. M. T., Hughes, C. E. (2008). Diversity and evolutionary history of lupins–insights from new phylogenies. In *Lupins for Health and Wealth*; Palta, J. A., Berger, J. B., Eds.; International Lupin Association: Canterbury, New Zealand. 346–354.

Ganesan, K., & Xu, B. (2017). Polyphenol-rich lentils and their health promoting effects. *International Journal of Molecular Sciences*, 18(11), 2390. doi:10.3390/ijms18112390.

Ha, P. T. H., & Luan, N. T. (2022). Nutraceutical Properties of Legume Seeds: Phytochemical Compounds. IntechOpen. *Legumes Research* doi: 10.5772/intechopen.100171. doi:10.5772/intechopen.100171.

Hernández-Morales, R., Ponce de León, S., Cruz-González, A., Aguilar-Cámara., Álvarez-Carreño, C., Becerra, A., & Lazcano, A. (2020). On the *Significance of Antiviral Combinations to Treat COVID-19:* Polymerase Plus Protease Inhibitors. Available at SSRN: https://ssrn.com/abstract=3653571 or doi:10.2139/ssrn.3653571.

Jia, X., Chen, M., Wan, J. B., Su, H., He, C. (2015). Review on the extraction, characterization and application of soybean polysaccharide. *RSC Advances*. 5:73525–34. doi: 10.1039/C5RA12421B.

Jin, A. L., Ozga, J. A., Lopes-Lutz, D., Schieber, A., & Reinecke, D. M. (2012). Characterization of proanthocyanidins in pea (*Pisum sativum* L.), lentil (*Lens culinaris* L.), and faba bean (*Vicia faba* L.) seeds. *Food Research International*, *46*(2), 528-535. doi: 10.1016/j.foodres.2011.11.018.

Kang, H. W. (2015). Antioxidant activity of ethanol and water extracts from lentil (Lens culinaris). *Journal of Food and Nutrition Research*, *3*(10), 667-669. doi:10.12691/jfnr-3-10-8.

Karamać M, Orak HH, Amarowicz R, Orak A, Piekoszewski W. Phenolic contents and antioxidant capacities of wild and cultivated white lupin (Lupinus albus L.) seeds. *Food Chemistry*. 2018 Aug 30; 258:1-7. doi: 10.1016/j.foodchem.2018.03.041.

Karataş, S. Ç., Günay, D., & Sayar, S. (2017). In vitro evaluation of whole faba bean and its seed coat as a potential source of functional food components. *Food Chemistry, 230*, 182-188. doi:10.1016/j.foodchem.2017.03.037.

Keogh J., Atkinson F., Eisenhauer B., Inamdar A. and Brand-Miller J. (2011). Food intake, postprandial glucose, insulin and subjective satiety responses to three different bread-based test meals. *Appetite*.57:707–710. doi: 10.1016/j.appet.2011.08.015.

Khalil, M. I., Salih, M. A., & Mustafa, A. A. (2020). Broad beans (*Vicia faba*) and the potential to protect from COVID-19 coronavirus infection. *Sudanese Journal of Paediatrics, 20*(1), 10. doi: 10.24911/SJP.106-1585398078.

Kouris, B. A., & Belski, R. (2016). Health benefits of legumes and pulses with a focus on Australian sweet lupins. *Asia Pacific Journal of Clinical Nutrition, 25*(1), 1-17. doi: 10.6133/apjcn.2016.25.1.23.

Kripi, V., Vandana, G., and Harish, D. (2019). Ethnopharmacology, Phytochemistry and Pharmacology of *Lens culinaris* Medikus Seeds: An Update. *Current Nutrition & Food Science*, 15(2), 121-129. doi: 10.2174/1573401313666170925155508.

Kumar, Y., Basu, S., Goswami, D., Devi, M., Shivhare, U. S., & Vishwakarma, R. K. (2022). Anti-nutritional compounds in pulses: Implications and alleviation methods. *Legume Science, 4*(2), e111. doi:10.1002/leg3.111.

Lemus, C. A., Rivero, P. F., Monserrar, P. S., and Millan, L. M. C. (2023). Nutritional composition and biological activity of narrow-leafed lupins (*Lupinus angustifolius L.*) hydrolysates and seeds. *Food Chemistry*. 136104. doi:10.1016/j.foodchem. 2023.136104.

León-Espinosa, E. B., Sánchez-Chino, X., Garduño-Siciliano, L., Álvarez-González, R. I., Dávila-Ortiz, G., Madrigal-Bujaidar, E., ... & Jiménez-Martínez, C. (2016). Hypocholesterolemic and anticarcinogenic effect of *Vicia faba* protein hydrolyzates. *Nutrition and Cancer, 68*(5), 856-864. doi:10.1080/01635581.2016.1180406.

Li, H., Zou, L., Li, X. Y., Wu, D. T., Liu, H. Y., Li, H. B., & Gan, R. Y. (2022). Adzuki bean (Vigna angularis): Chemical compositions, physicochemical properties, health benefits, and food applications. *Comprehensive Reviews in Food Science and Food Safety, 21*(3), 2335-2362.doi: 10.1080/01635581.2016.1180406.

Li X, Xin Y, Mo Y, Marozik P, He T, Guo H. (2022). The Bioavailability and Biological Activities of Phytosterols as Modulators of Cholesterol Metabolism. *Molecules*. 27(2):523. doi:10.3390/molecules27020523.

Lima, N. M., Santos, V. N. C., & La Porta, F. D. A. (2018). Quimiodiversidade, bioatividade e quimiossistemática do gênero *Inga* (Fabaceae): uma breve revisão. *Revista Virtual Química*, 10(3), 459-473.

Liu, H., Xu, J., Xu, X., Yuan, Z., Song, H., Yang, L. (2021). Structure/function relationships of bean polysaccharides: a review. *Critical Reviews in Food Science*. 1−15. doi:_10.1080/10408398.2021.1946480.

Liu, Y., Tong, T, Han, R., Zhang, Y. Li, F, Shi, H. and Jiao, Y. (2023). Effect of different arrangements of globe partiles in radio frequency heating uniformity: Using black papper as an example. *LWT- Food Science and Technology*. 174: 114422. doi:10.1016/j.lwt.2022.114422.

Lopez, M. L. X. (2020). Bioactive compounds of runner bean (*Phaseolus coccineus* L.). *Bioactive Compounds in Underutilized Vegetables and Legumes*, 1-17.doi:10.1007/978-3-030-57415-4_31.

Looi, E. P., & MohdMaidin, N. (2023). The bioactivities of legumes: a review. *Food Research,* 7(5), 339-360. doi:10.5935/abc.20170158.

Makhmudova, U., Schulze, P. C., Lütjohann, D., & Weingärtner, O. (2021). Phytosterols and cardiovascular disease. *Current Atherosclerosis Reports*, 23(11), 68.doi:10.1007/s11883-021-00964-x.

Martineau, C. D., Achouri, A., Karboune, S., & L'Hocine, L. (2022). Faba bean: an untapped source of quality plant proteins and bioactives. *Nutrients*, 14(8), 1541. doi:10.3390/nu14081541.

Marventano, S., Pulido, M. I., Sánchez-González, C., Godos, J., Speciani, A., Galvano, F., & Grosso, G. (2017). Legume consumption and CVD risk: a systematic review and meta-analysis. *Public Health Nutrition*, 20(2), 245-254. doi: 10.29219/fnr.v67.9541.

Abdallah, E. M., Qureshi, K. A., & Musa, K. H. (2017). Antimicrobial, antioxidant and phytochemical screening of Lupin seeds (Lupinus termis Forrsk.) from Sudan. *CIBTech Journal of Microbiology*, *6*, 1-8.

Mohan V, Tresina P, Daffodil E. (2016) Antinutritional factors in legume seeds: characteristics and determination. *Encyclopedia of Food and Health.* 211-220. doi: 10.1016/B978-0-12-384947-2.00036-2.

Montserrat, P. S., Lemus, C. A., Toscano, R., Pedroche, J., Millan, F., and Millan, L. M. C. (2019). GPETAFLR, an octapeptide isolated from *Lupinus angustifolius* L. protein hydrolysate, promotes the skewing to the M2 phenotype in human primary monocytes. *Food & Function, 10*(6), 3303-3311. https://doi.org/10.1039/C9FO00115H.

Moreno, V. C. A., Luna, V. D., Camacho, R. R. M., & Mojica, L. (2020). Bioactive proteins and phytochemicals from legumes: Mechanisms of action preventing obesity and type-2 diabetes. *Food Research International*, 130, 108905. doi:10.1016/j.foodres. 2019.108905.

Méndez, L. L. F., Sosa, L. D., López, C., Lomelí, M., González, M. B. E., and Vázquez, R. J. A. (2022). Phytochemicals From *Vicia faba* Beans as Ligands of the Aryl Hydrocarbon Receptor to Regulate Autoimmune Diseases. *Frontiers in Nutrition.* 9:790440. doi: 10.3389/fnut.2022.790440.

Muñoz, E. B., Luna, V. D. A., Fornasini, M., Baldeón, M. E., & de Mejia, E. G. (2018). Gamma-conglutin peptides from Andean lupin legume (*Lupinus mutabilis* Sweet) enhanced glucose uptake and reduced gluconeogenesis *in vitro*. *Journal of Functional Foods,* 45, 339-347.doi:10.1016/j.jff.2018.04.021.

Pan, W. L., Ng, T. B. (2015). A dimeric *Phaseolus coccineus* lectin with anti-oxidative, antiproliferative and cytokine-inducing activities. *International Journal of Biological Macromolecules* 81:960–966. doi: 10.1016/j.ijbiomac.2015.09.034.

Pereira, A., Ramos, F., & Sanches, S. A. (2022). Lupin (*Lupinus albus* L.) seeds: Balancing the good and the bad and addressing future challenges. *Molecules*, 27(23), 8557.doi: 10.3390/molecules27238557.

Ros E, Hu FB. (2013). Consumption of plant seeds and cardiovascular health: epidemiological and clinical trial evidence. *Circulation.* 128(5):553-65. doi: 10.1161/CIRCULATIONAHA.112.001119.

Ruiz-López, M. A., Barrientos-Ramírez, L., García-López, P. M., Valdés-Miramontes, E. H., Zamora-Natera, J. F., Rodríguez-Macias, R., & Vargas-Radillo, J. J. (2019). Nutritional and bioactive compounds in mexican lupin beans species: A mini-review. *Nutrients*, *11*(8), 1785. doi:10.3390/nu11081785.

Santos, S. G., Cruz, C. I., Álvarez, R. A. I., Álvarez, S. N., Rodríguez, O. B., Álvarez, L. A. I., & Carrillo, V. A. (2022). Bioactive Peptides from Lupin (*Lupinus angustifolius*) Prevent the Early Stages of Atherosclerosis in Western Diet-Fed ApoE–/–Mice. *Journal of Agricultural and Food Chemistry*, 70(27), 8243-8253.doi: 10.1021/acs.jafc.2c00809.

Sedláková, K., Straková, E., Suchý, P., Krejcarová, J., & Herzig, I. (2016). Lupin as a perspective protein plant for animal and human nutrition–a review. *Acta Veterinaria Brno*, *85*(2), 165-175.doi: 10.2754/avb201685020165.

Serrano-Sandoval, S. N.; Guardado-Felix, D.; Gutiérrez-Uribe, J. A. Legumes in Human Health and Nutrition. Encyclopedia of Human Nutrition: Volume 1-4, Fourth Edition 2023, 1–4, 430–437, doi:10.1016/B978- 0-12-821848-8.00061-5.

Shimada, K., Fujikawa, K., Yahara, K., & Nakamura, T. (1992). Antioxidative properties of xanthan on the autoxidation of soybean oil in cyclodextrin emulsion. *Journal of Agricultural and Food Chemistry*, *40*(6), 945-948. doi:10.1021/jf00018a005.

Siah, S. D., Konczak, I., Agboola, S., Wood, J. A., & Blanchard, C. L. (2012). *In vitro* investigations of the potential health benefits of Australian-grown faba beans (*Vicia faba* L.): chemopreventative capacity and inhibitory effects on the angiotensin-converting enzyme, α-glucosidase and lipase. *British Journal of Nutrition*, *108*(S1), S123-S134.doi:10.1017/S0007114512000803.

Singh, B., Singh, J. P., Shevkani, K., Singh, N., & Kaur, A. (2017). Bioactive constituents in pulses and their health benefits. *Journal of Food Science and Technology*, 54, 858-870. doi:10.1007%2Fs13197-016-2391-9.

Singhal, P., Kaushik, G., Mathur, P. (2014). Antidiabetic potential of commonly consumed legumes: A review. *Critical Reviews in Food Science and Nutrition*. 54(5):655-672. doi: 10.1080/10408398.2011.604141.

Turco, I., Ferretti, G., & Bacchetti, T. (2016). Review of the health benefits of Faba bean (Vicia faba L.) polyphenols. *Journal of Food & Nutrition Research*, *55*(4).

Alam, T., Najam, L. (2022). Faba-Bean Antioxidant and Bioactive Composition: Biochemistry and Functionality. In: Punia Bangar, S., Bala Dhull, S. (eds) Faba Bean: Chemistry, Properties and Functionality. Springer, Cham. doi:10.1007/978-3-031-14587-2_6.

M. van de Noort, Chapter 11 - Lupin: An Important Protein and Nutrient Source, Editor(s): Sudarshan Nadathur, Janitha P. D. Wanasundara, Laurie Scanlin, Sustainable Protein Sources (Second Edition), Academic Press, 2024, Pages 219-239, ISBN 9780323916523, doi:.org/10.1016/B978-0-323-91652-3.00028-9.

Villarino, C. B. J. (2014). *Maximising the nutritional and sensory quality of lupin bread made using western Australian bakers flour* (Doctoral dissertation, Curtin University).

Villarino, C. B. J., Jayasena, V., Coorey, R., Chakrabarti-Bell, S., & Johnson, S. K. (2016). Nutritional, health, and technological functionality of lupin flour addition to bread and other baked products: Benefits and challenges. *Critical Reviews in Food Science and Nutrition*, *56*(5), 835-857. doi: 10.1080/10408398.2013.814044.

Wang, W., Li, Q., Wu, J., Hu, Y., Wu, G., Yu, C., ... & Wang, Y. (2021). Lentil lectin derived from *Lens culinaris* exhibit broad antiviral activities against SARS-CoV-2 variants. *Emerging microbes & infections*, *10*(1), 1519-1529. doi: 10.1080/22221751.2021.1957720.

Zhang, B., Deng, Z., Ramdath, D. D., Tang, Y., Chen, P. X., Liu, R., & Tsao, R. (2015). Phenolic profiles of 20 Canadian lentil cultivars and their contribution to antioxidant activity and inhibitory effects on α-glucosidase and pancreatic lipase. *Food Chemistry*, *172*, 862-872. doi: 10.1016/j.foodchem.2014.09.144.

# Chapter 13

# Functional Potential of Non-Conventional Legumes

**C. Jiménez-Martínez[1]**
**V. Pérez-Pérez[2]**
**M. F. Juárez-Chairez[3]**
**D. R. Moguel-Concha[1]**
**J. E. Borges-Martínez[1]**
**A. E. Cedillo-Olivos[1]**
**and M. S. Cid-Gallegos[2,*]**

[1]Escuela Nacional de Ciencias Biológicas, Instituto Politécnico Nacional, Unidad Profesional Adolfo López Mateos, Mexico City, Mexico
[2]Unidad Profesional Interdisciplinaria de Biotecnología, Instituto Politécnico Nacional, México City, Mexico
[3]Laboratorios de Inmunología y Biología Molecular, Unidad Académica de Ciencias Biológicas de la Universidad Autónoma de Zacatecas, Guadalupe, Zacatecas, México

## Abstract

A wide variety of seeds are consumed globally as part of the human diet. Among these, common legumes such as beans, soybeans, and chickpeas are the most widely used. However, there are also pulses consumed locally, referred to as non-conventional legumes because they are not commonly utilized. These underutilized pulses are increasingly recognized as a valuable resource for enhancing food security and providing plant-based proteins to the global population. Their appeal

* Corresponding Author's Email: cid.stephanie@gmail.com

In: Legumes: Beyond Their Nutritional Value
Editors: C. Jiménez-Martínez, M. F. Juárez-Chairez and M. S. Cid-Gallegos
ISBN: 979-8-89530-440-2

extends beyond their nutritional value, as they contain bioactive compounds with potential applications in the food, pharmaceutical, and cosmetic industries. As a result, the use of non-conventional legumes is gaining momentum, driven by the health benefits conferred by their bioactive compounds. This chapter will describe some of the biofunctional properties of compounds present in *Vigna unguiculata, Pithecellobium, Leucaena, Hymenaea courbaril, Canavalia ensiformis* are described.

**Keywords:** non-conventional legumes, bioactivity, techno-functional, food industry

## 1. Introduction

In recent years, the nutritional and bioactive potential of non-conventional legumes has garnered increasing global interest, despite their current use being largely limited to their regions of origin. However, these legumes are not widely adopted worldwide due to agronomic, economic, and cultural factors, as more commonly used pulses such as beans, soybeans, and chickpeas dominate in many regions and applications (Cheng et al., 2019). Therefore, promoting these legume species is necessary as they represent a plant-based protein food source due to their adequate concentration of proteins and essential nutrients (Henchion et al., 2017). In addition, the amount of protein in these legumes is like that of those most consumed. They also provide a large amount of structural carbohydrates due to their high contribution of dietary fiber, which can lower serum cholesterol levels due to their low glycemic index (Martín-Cabrejas et al., 2008). Additionally, non-conventional legumes have many non-nutritional compounds that may have adverse effects, but they present biological activity important for human health in low concentrations (Alcázar-Valle et al., 2020; Carbas et al., 2020; Samtiya et al., 2020). The adverse effects of anti-nutritional factors such as saponins, phytates, and trypsin inhibitors, among others, are mainly the inhibition or absorption of proteins and certain minerals (Fe and Zn) (Cid-Gallegos et al., 2022; Gupta et al., 2015; Samtiya et al., 2020). On the other hand, it has been reported that non-nutritional compounds have shown antioxidant, antimicrobial, anti-inflammatory, angiotensin I-converting enzyme inhibitor activity, among others (Harvian et al., 2019; Jang, 2023; Khanzada et al., 2013; Spera et al., 2019). Moreover, some biological activities of non-conventional legumes

(*Vigna unguiculata, Pithecellobium, Hymenaea, Canavalia ensiformis)* (Figure 1) will be described in the present chapter.

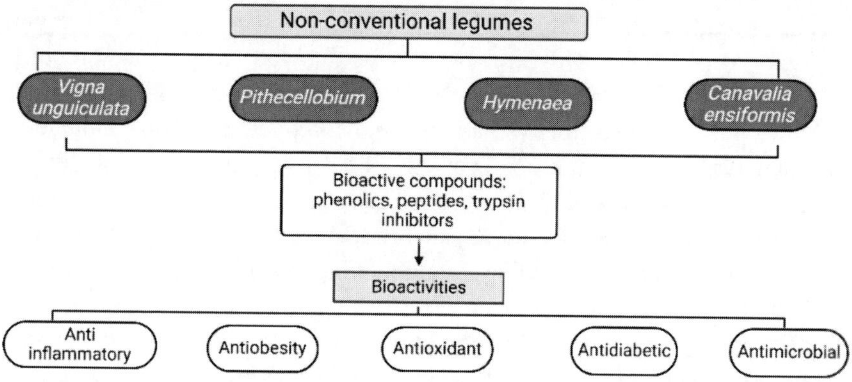


**Figure 1.** Non-conventional legumes and their bioactivities. Created in Bio-render.

## 2. *Vigna unguiculata*

### 2.1. Bioactive Compounds

*Vigna unguiculata* grows worldwide and is a good source of proteins (Gupta et al., 2016). On the other side, they have a great content of compounds with biological activity that benefit the consumer's health, such as antioxidants (Tzanova et al., 2023). Some bioactive compounds studied are phenolics, and peptides, among others. Even though *Vigna* is an underutilized legume, it is used in traditional medicine to treat neuritis, insomnia, weakness of memory, dyspepsia, indigestion, needles in limbs, and the sensation of pins (Zaheer et al., 2020). Additionally, phenolic compounds are found in legume plants and are used in diverse areas as industrial chemicals, agrochemicals, pharmaceuticals, and other consumer products because of the biological characteristics to chelate metals, modulate enzyme activity, combat free radicals (Santos et al., 2021).

### 2.2. Bio-Functional Properties

Some of the biological activities reported for cowpea are anti-diabetic, anti-inflammatory, antioxidant, and antimicrobial, in particular (Table 1) (Tzanova et al., 2023; Moloto et al., 2020; Ashraduzzaman et al., 2016; López-Morales et al., 2020). Cowpea consumption has been associated with a decreased risk of type 2 diabetes mellitus and obesity (Priyadarshini et al., 2024). Moloto et al. (2020) studied the inhibition of α-amylase, α-glucosidase and the gene expression of GLUT-4 with extracts of seven cowpea leaves; they reported that the extract of cowpea cultivar VOP1 at 25 and 50 mg/mL show higher inhibitory percentages values (86 and 93%). Moreover, Ashraduzzaman et al. (2016) analyzed the antimicrobial activity of three *Vigna unguiculata* seed oil varieties and mentioned all three oils were active against the three tested fungi. Other research evaluated the anti-neurodegenerative effect of phenolic extracts from pods and seeds; they found that higher bioactivity was observed in the pod extracts (Avanza et al., 2021). In addition, Priyadarshini et al. (2024) studied aqueous and ethanolic extracts from seeds and reported DPPH scavenging, α-amylase inhibitory activity, and anti-inflammatory potential. Chidebelu et al. (2019) evaluated ethanolic extracts from seeds and found that only 50% of the strains were inhibited by *Vigna unguiculata.* Furthermore, Gutiérrez-Uribe et al. (2011) analyzed extracts from whole cowpeas, seed coats, and cotyledons on MCF-7 cells and mentioned that free phenolic extract of whole seeds inhibited 65% of the proliferation. Similarly, research of seed extracts from *Vigna unguiculata* demonstrated the ability to reduce the level of Epidermal Growth Factor Receptor (EGFR) phosphorylation in E705, DiFi, and SW480 cell lines (Panzeri et al., 2020). Additionally, Sandeep (2014) worked with aqueous and ethanolic extracts of seeds and reported antibacterial activity against Gram-positive bacteria, *Bacillus subtilis,* and Gram-negative bacteria, *Escherichia coli.* Thumbrain et al. (2020) studied protein isolates from five cowpea cultivars and demonstrated cowpea isolates show high potential to use as therapeutics in cancer.

## 3. *Pithecellobium*

### 3.1. Bioactive Compounds

Secondary metabolites such as flavonoids, anthocyanins, and terpenoids derived from plants could exert diverse pharmacological activities like antifungal, antioxidant, antibacterial, and anti-inflammatory (Pagare et al., 2015). *Pithecellobium* is reportedly used in folk medicine for earache, leprosy, peptic ulcer, toothache, emollient, anodyne, and larvicidal. Specifically, parts like the skin of the stem for dysentery, leaves for intestinal disorders, and seeds for ulcers, for example (Kumar et al., 2013). Moreover, some biological activities with beneficial effects on health have been reported, such as anti-inflammatory, anti-venom, protease inhibitory, spermicidal, antimicrobial, and antitubercular activity (Kumar et al., 2013).

### 3.2. Bio-Functional Properties

*Pithecellobium* is an underutilized legume, traditionally consumed regionally. This legume possesses numerous biological properties, which are summarized in Table 1. Kumar et al. (2013) analyzed leaf extracts of *Pithecellobium* against twenty pathogenic microorganisms and reported that the extracts inhibited the growth of most of these microbes. In addition, Katekhaye & Kale (2012) revised extracts of wood bark and leaves to evaluate their antioxidant activity and reported that the extracts were suitable iron chelators. Research with seed extracts of *Pithecellobium* showed antibacterial activity towards tested clinical bacterial strains (Aldarhami et al., 2023). Moreover, Kalavani et al. (2016) studied ethanolic extracts and demonstrated antibacterial activities against four bacteria and anti-inflammatory responses, such as the percentage of inhibition of protein denaturation. Furthermore, Saeed Kotb et al. (2020) showed antioxidant activity with bark and leaf extracts of *Pithecellobium.*

## 4. *Hymenaea*

### 4.1. Bioactive Compounds

Since ancient times, the *Hymenaea* tree has been used in folk medicine as an anti-inflammatory agent, appetite stimulant, and energy tonic. In addition, the bark is used to treat anemia and coughs (Silva et al., 2015). Secondary metabolites present in *Hymenaea*, like flavonoids, tannins, and terpenes, exert antioxidant activity attributed to their chemical structure and reducing properties (Santos et al., 2017; Brunelli et al., 2014).

**Table 1.** Bio-functional activities of non-conventional legumes

| Non- conventional legumes | Used Part | Biological effect | Reference |
|---|---|---|---|
| *Vigna unguiculata* | Leaves | Antidiabetic | Moloto et al., 2020 |
| | Oil | Antimicrobial | Ashraduzzaman et al., 2016 |
| | Pods and seeds | Anti-neurodegenerative | Avanza et al., 2021 |
| | Seeds | Antidiabetic<br>Antioxidant<br>Anti-inflammatory | Priyadarshini et al., 2024 |
| | Seeds | Antimicrobial | Chidebelu et al., 2019 |
| | Seeds<br>Testa<br>Cotyledons | Cancer cell inhibition (MCF-7) | Gutierrez-Uribe et al., 2011 |
| | Seeds | Colorectal cancer prevention | Panzeri et al., 2020 |
| | Seeds | Antibacterial | Sandeep, 2014 |
| | Protein isolates from seeds | Antioxidant<br>Apoptotic | Thumbrain et al., 2020 |
| *Pithecellobium* | Leaves | Antimicrobial | Kumar et al., 2013 |
| | Wood bark and leaves | Antioxidant | Katekhaye & Kale, 2012 |
| | Seed | Antimicrobial | Aldarhami et al., 2023 |
| | Seed | Anti-inflammatory<br>Antibacterial | Kalavani et al., 2016 |
| | Bark and leaves | Antioxidant | Saeed Kotb et al., 2020 |
| *Hymenaea* | Pulp | Antioxidant | Pereira Santos et al., 2020 |
| | Sap | Antioxidant | Da Costa et al., 2021 |
| *Canavalia ensiformis* | Seeds | Antioxidant<br>Antidiabetic | Vadivel et al., 2012 |
| | Hydrolysates | Antioxidant | Wijatniko & Murdiati, 2019 |

## 4.2. Biological Activities

*Hymenaea* is a non-conventional legume with many bioactive compounds that must be studied. Table 1 will show some research on bioactive compounds present in *Hymenaea.* Pereira Santos et al. (2020) investigated pulp and polyphenols and reported antioxidant capacity. Finally, da Costa et al. (2021) evaluated the sap and mentioned it has antioxidant activity, inhibiting the production of reactive oxygen species.

## 5. *Canavalia ensiformis*

### 5.1. Bioactive Compounds

*Canavalia ensiformis* is an underutilized legume with natural phenolic antioxidants that can scavenge reactive oxygen and nitrogen species (Doss et al., 2011).

### 5.2. Biological Activities

*Canavalia ensiformis* is an underutilized legume with many bioactive compounds that must be investigated to upgrade the available information. Table 1 will show some research on bioactive compounds present in *Canavalia ensiformis*. Vadivel et al. (2012) studied a methanolic extract and reported antioxidant and antidiabetic activity by inhibition of α-amylase and α-glucosidase. Wijatniko & Murdiati (2019) analyzed hydrolysates from jack beans, and the hydrolysate with pepsin showed higher DPPH scavenging.

## Conclusion

Underutilized or regionally consumed legumes deserve greater attention from the research community, as they represent a valuable source of bioactive compounds with potential applications across various industries.

## References

Alcázar-Valle, M., Lugo-Cervantes, E., Mojica, L., Morales-Hernández, N., Reyes-Ramírez, H., Enríquez-Vara, Navat, J., & García-Morales, S. (2020). Bioactive Compounds, Antioxidant Activity, and Antinutritional Content of Legumes: A

Comparison between Four *Phaseolus* Species. *Molecules (Basel, Switzerland)*, *25*(15), 3528. https://doi.org/10.3390/molecules25153528.

Aldarhami, A., Bazaid, A. S., Alhamed, A. S., Alghaith, A. F., Ahamad, S. R., Alassmrry, Y. A., Alharazi, T., Snoussi, M., Qanash, H., Alamri, A., Badraoui, R., Kadri, A., Binsaleh, N. K., & Alreshidi, M. (2023). Antimicrobial Potential of *Pithecellobium dulce* Seed Extract against Pathogenic Bacteria: In silico and in vitro Evaluation. *BioMed research international*, *2023*, 2848198. https://doi.org/10.1155/2023/2848198.

Ashraduzzaman, M., Alam, M. A., Khatun, S., & Absar, N. (2016). Antimicrobial activity of *Vigna unguiculata* L. Walp seed oil. *International Journal of Biotechnology*, *5*(3), 71.

Avanza, M. V., Álvarez-Rivera, G., Cifuentes, A., Mendiola, J. A., & Ibáñez, E. (2021). Phytochemical and Functional Characterization of Phenolic Compounds from Cowpea (*Vigna unguiculata* (L.) Walp.) Obtained by Green Extraction Technologies. *Agronomy*, *11*(1).

Brunelli, D. T., Caram, K., Nogueira, F. R., Libardi, C. A., Prestes, J., & Cavaglieri, C. R. (2014). Immune responses to an upper body tri-set resistance training session. *Clinical Physiology and Functional Imaging*, *34*(1), 64-71. https://doi.org/https://doi.org/10.1111/cpf.12066.

Carbas, B., Machado, N., Oppolzer, D., Ferreira, L., Queiroz, M., Brites, C., Rosa, E. A., & Barros, A. I. (2020). Nutrients, Antinutrients, Phenolic Composition, and Antioxidant Activity of Common Bean Cultivars and their Potential for Food Applications. *Antioxidants (Basel)*, *9*(2). https://doi.org/10.3390/antiox9020186.

Cheng, A., Raai, M. N., Zain, N. A. M., Massawe, F., Singh, A., & Wan-Mohtar, W. A. A. Q. I. (2019). In search of alternative proteins: unlocking the potential of underutilized tropical legumes. *Food Security*, *11*(6), 1205-1215. https://doi.org/10.1007/s12571-019-00977-0.

Chidebelu, P., Onovo, I., & Nweze, E. I. (2019). Antimicrobial activities of *Cajanus cajan* L., *Phaseolus vulgaris* L. and *Vigna unguiculata* L. against some bacterial and fungal isolates. *Journal of Basic Pharmacology and Toxicology*, *3*(1), 29-33. http://scigreen.com/index.php/JBPT/article/view/58.

Cid-Gallegos, M. S., Corzo-Ríos, L. J., Jiménez-Martínez, C., & Sánchez-Chino, X. M. (2022). Protease Inhibitors from Plants as Therapeutic Agents- A Review. *Plant Foods for Human Nutrition*. https://doi.org/10.1007/s11130-022-00949-4.

da Costa, R. S. L., Martins, D. S., Peixoto, L. D. C., da Silva, B. J. P., Borges, L. B., Lima, E. S., Koolen, H. H. F., de Oliveira Meneguetti, D. U., Pessoa, A. F. M., & Silva, R. P. M. (2021). Antioxidant effect of *Hymenaea courbaril* L (Jatob) sap on the healing of wounds on mice. *Journal of Medicinal Plants Research*, *15*(4), 160-171.

Doss, A., Pugalenthi, M., & Vadivel, V. (2011). Antioxidant activity of raw and differentially processed under-utilized tropical legume *Canavalia ensiformis* L. DC seeds, South India. *The IIOAB Journal*, *2*(8), 27-32.

Gupta, K., Das, S., Aman, S., & Nayak, A. (2016). Pharmacological activities of *Vigna unguiculata*. *World Journal of Pharmaceutical Research*, *5*(10), 337-345.

Gupta, R. K., Gangoliya, S. S., & Singh, N. K. (2015). Reduction of phytic acid and enhancement of bioavailable micronutrients in food grains. *Journal of food science and technology*, *52*(2), 676-684. https://doi.org/10.1007/s13197-013-0978-y.

Gutiérrez-Uribe, J. A., Romo-Lopez, I., & Serna-Saldívar, S. O. (2011). Phenolic composition and mammary cancer cell inhibition of extracts of whole cowpeas (*Vigna unguiculata*) and its anatomical parts. *Journal of Functional Foods*, *3*(4), 290-297. https://doi.org/https://doi.org/10.1016/j.jff.2011.05.004.

Harvian, Z. A., Ningrum, A., Anggrahini, S., & Setyaningsih, W. (2019). In silico approach in evaluation of jack bean (*Canavalia ensiformis*) canavalin protein as precursors of bioactive peptides with dual antioxidant and angiotensin i-converting enzyme inhibitor. Materials Science Forum.

Henchion, M., Hayes, M., Mullen, A. M., Fenelon, M., & Tiwari, B. (2017). Future Protein Supply and Demand: Strategies and Factors Influencing a Sustainable Equilibrium. *Foods*, *6*(7), 53. https://www.mdpi.com/2304-8158/6/7/53.

Jang, H.-L. (2023). Effect of sword bean (*Canavalia ensiformis*) fermentation filtrate on the antioxidant, anti-inflammatory, and antimicrobial activities. *Korean Journal of Food Preservation*, *30*(6), 1072-1081.

Kalavani, R., Banu, R. S., Jeyanthi, K., Sankari, T. U., & Kanna, A. V. (2016). Evaluation of anti-inflammatory and antibacterial activity of *Pithecellobium dulce* (Benth) extract. *Biotechnological Research*, *2*(4), 148-154.

Katekhaye, S. D., & Kale, M. S. (2012). Antioxidant and free radical scavenging activity of *Pithecellobium dulce* (Roxb.) Benth wood bark and leaves. *Free Radicals and Antioxidants*, *2*(3), 47-57. https://doi.org/https://doi.org/10.5530/ax.2012.3.7.

Khanzada, S. K., Khanzada, A. K., Shaikh, W., & Ali, S. A. (2013). Phytochemical studies on *Pithecellobium dulce* Benth. A medicinal plant of Sindh, Pakistan. *Pakistan Journal of Botany*, *45*(2), 557-561.

Kumar, M., Nehra, K., & Duhan, J. (2013). Phytochemical analysis and antimicrobial efficacy of leaf extracts of *Pithecellobium dulce*. *Asian Journal of Pharmaceutical and Clinical Research*, *6*(1), 70-76.

López-Morales, D., de la Cruz-Lázaro, E., Sánchez-Chávez, E., Preciado-Rangel, P., Márquez-Quiroz, C., & Osorio-Osorio, R. (2020). Impact of Agronomic Biofortification with Zinc on the Nutrient Content, Bioactive Compounds, and Antioxidant Capacity of Cowpea Bean (*Vigna unguiculata* L. Walpers). *Agronomy*, *10*(10).

Martín-Cabrejas, M. A., Díaz, M. F., Aguilera, Y., Benítez, V., Mollá, E., & Esteban, R. M. (2008). Influence of germination on the soluble carbohydrates and dietary fibre fractions in non-conventional legumes. *Food Chemistry*, *107*(3), 1045-1052. https://doi.org/https://doi.org/10.1016/j.foodchem.2007.09.020.

Moloto, M. R., Phan, A. D. T., Shai, J. L., Sultanbawa, Y., & Sivakumar, D. (2020). Comparison of Phenolic Compounds, Carotenoids, Amino Acid Composition, In Vitro Antioxidant and Anti-Diabetic Activities in the Leaves of Seven Cowpea (*Vigna unguiculata*) Cultivars. *Foods*, *9*(9).

Pagare, S., Bhatia, M., Tripathi, N., Pagare, S., & Bansal, Y. (2015). Secondary metabolites of plants and their role: Overview. *Current trends in biotechnology and pharmacy*, *9*(3), 293-304.

Panzeri, D., Guzzetti, L., Sacco, G., Tedeschi, G., Nonnis, S., Airoldi, C., Labra, M., Fusi, P., Forcella, M., & Regonesi, M. E. (2020). Effectiveness of *Vigna unguiculata* seed extracts in preventing colorectal cancer [10.1039/D0FO00913J]. *Food & function, 11*(7), 5853-5865. https://doi.org/10.1039/D0FO00913J.

Pereira Santos, A. C., Martins Alves, A., Veloso Naves, M. M., & Reis Silva, M. (2020). Nutritional profile, bioactive compounds and antioxidant capacity of jatobá-da-mata (*Hymenaea courbaril*, var. stilbocarpa) by product. *Revista Chilena de Nutrición*, 366-371. https://doi.org/http://dx.doi.org/10.4067/S0717-75182020000300366.

Priyadarshini, R. D., Beatrice, D. A., & Sivaraj, C. (2024). Antioxidant Potential, Antidiabetic, and Anti-Inflammatory Activities of White Cowpea (*Vigna unguiculata* L.): An in vitro Study. *Asian Journal of Biological and Life Sciences*, *12*(3).

Saeed Kotb, S., Ayoub, I. M., El-Moghazy, S. A., & Singab, A. N. B. (2020). Profiling the Lipophilic Fractions of *Pithecellobium dulce* Bark and Leaves Using GC/MS and Evaluation of Their Antioxidant, Antimicrobial and Cytotoxic Activities. *Chemistry & Biodiversity*, *17*(7), e2000048. https://doi.org/https://doi.org/10.1002/cbdv.202000048.

Samtiya, M., Aluko, R. E., & Dhewa, T. (2020). Plant food anti-nutritional factors and their reduction strategies: an overview. *Food Production, Processing and Nutrition*, *2*(1), 6. https://doi.org/10.1186/s43014-020-0020-5.

Sandeep, D. (2014). Evaluation of antibacterial activity of seed extracts of *Vigna unguiculata*. *International Journal of Pharmacy and Pharmaceutical Sciences*, *6*(1), 75-77.

Santos, R. A., Souza Filho, A. P. S., Cantanhede Filho, A. J., Guilhon, G., & Santos, L. S. (2021). Analysis of phenolic compounds from cowpea (*Vigna unguiculata*) by HPLC-DAD-MS/MS. *Brazilian Journal of Food Technology*, *24*, e2020077. https://doi.org/https://doi.org/10.1590/1981-6723.07720.

Santos, W. N. L. d., da Silva Sauthier, M. C., dos Santos, A. M. P., de Andrade Santana, D., Almeida Azevedo, R. S., & da Cruz Caldas, J. (2017). Simultaneous determination of 13 phenolic bioactive compounds in guava (*Psidium guajava* L.) by HPLC-PAD with evaluation using PCA and Neural Network Analysis (NNA). *Microchemical Journal*, *133*, 583-592. https://doi.org/https://doi.org/10.1016/j.microc.2017.04.029.

Silva, A., Rabelo, M., & Enoque, M. (2015). Diversidade de angiospermas e espécies medicinais de uma área de Cerrado. *Revista Brasileira de Plantas Medicinais*, *17*, 1016-1030.

Spera, K. D., Figueiredo, P. A., Santos, P. C., Barbosa, F. C., Alves, C. P., Dokkedal, A. L., Saldanha, L. L., Silva, L. P., Figueiredo, C. R., & Ferreira, P. C. (2019). Genotoxicity, anti-melanoma and antioxidant activities of *Hymenaea courbaril* L. seed extract. *Anais da Academia Brasileira de Ciências*, *91*.

Thumbrain, D., Dwarka, D., Gerrano, A. S., & Mellem, J. J. (2020). Antioxidant and apoptotic potential of protein isolates derived from *Vigna unguiculata* (L.) Walp. *International Journal of Food Science & Technology*, *55*(7), 2813-2823. https://doi.org/https://doi.org/10.1111/ijfs.14535.

Tzanova, M. T., Stoilova, T. D., Todorova, M. H., Memdueva, N. Y., Gerdzhikova, M. A., & Grozeva, N. H. (2023). Antioxidant Potentials of Different Genotypes of Cowpea

(*Vigna unguiculata* L. Walp.) Cultivated in Bulgaria, Southern Europe. *Agronomy, 13*(7).

Vadivel, V., Cheong, J. N., & Biesalski, H. K. (2012). Antioxidant and type II diabetes related enzyme inhibition properties of methanolic extract of an underutilized food legume, *Canavalia ensiformis* (L.) DC: Effect of traditional processing methods. *LWT, 47*(2), 255-260. https://doi.org/https://doi.org/10.1016/j.lwt.2012.01.014.

Wijatniko, B. D., & Murdiati, A. (2019). Antioxidant activity of bioactive peptides derived from the hydrolysates of jack bean (*Canavalia ensiformis* (L.) DC.) protein isolate. *AIP Conference Proceedings, 2099*(1), 020028. https://doi.org/10.1063/1.5098433.

Zaheer, M., Ahmed, S., & Hassan, M. M. (2020). *Vigna unguiculata* (L.) Walp. (Papilionaceae): A review of medicinal uses, Phytochemistry and pharmacology. *Journal of Pharmacognosy and phytochemistry, 9*(1), 1149-1152.

# Index

## D



## E



## F



## G

## H



## I



## K



## L



## M

## Q



## R



## S



## T

## U



## V



## W